The Science of *Breaking Bad*

The Science of *Breaking Bad*

Dave Trumbore and Donna J. Nelson

The MIT Press
Cambridge, Massachusetts
London, England

This book was set in ITC Stone Serif Std and ITC Stone Sans Std by Toppan Best-set Premedia Limited. Printed and bound in the United States of America.

Library of Congress Cataloging-in-Publication Data

Names: Trumbore, Dave, author. | Nelson, Donna J., author.
Title: The science of Breaking Bad / Dave Trumbore and Donna J. Nelson.
Description: Cambridge, MA : The MIT Press, [2019] | Includes bibliographical references and index.
Identifiers: LCCN 2018041803 | ISBN 9780262537155 (paperback : alk. paper)
Subjects: LCSH: Breaking bad (Television program : 2008-2013)
Classification: LCC PN1992.77.B74 T78 2019 | DDC 791.45/72--dc23 LC record available at https://lccn.loc.gov/2018041803

10 9 8 7 6 5 4 3 2 1

To Mo, thanks for everything.—So (Dave)

To the arm-chair scientists, who love *Breaking Bad* and made it one of the greatest television series of all time.—Dr. Donna J. Nelson

"This is a story about a man who transforms himself from Mr. Chips into Scarface."
Vince Gilligan

Contents

Foreword

Breaking Bad is not about meth; it's about greed. About the way greed transforms a humble chemistry teacher into a merciless drug lord. By watching the show, nobody will learn how to make methamphetamine. They might, however, learn a few things about human nature, the rough road to domination, and the destructive power of absolute power.

Yes, I just used "power" twice in a sentence but what do I know, I'm just a chemist. And so is Dr. Donna J. Nelson, a professor of chemistry at the University of Oklahoma and former president of the American Chemical Society. She was instrumental in helping the show's creative team present spectacular and slightly wrong methamphetamine synthesis processes. That's right, some of you might have figured out already (the hard way) that *Breaking Bad* is not a "how-to-make-meth" guide. And neither is *The Science of "Breaking Bad,"* the brainchild of talented pop science writer Dave Trumbore, who coauthored the book with Dr. Nelson.

The duo lead us through a catalog of science-related moments found throughout the five seasons of the series. And yes, there is a lot of chemistry in this book. The explanations come in both simple terms (section 101 in each chapter) and more advanced descriptions. The reader also gets bonus content such as trivia and side reactions ("Side RxN")—worthy scientific observations extracted from various episodes.

Even addicted viewers will discover something interesting. Take the original opening credits as an example. Many have figured out that 35 and 56 are the atomic numbers of bromine (Br) and barium (Ba), respectively, and that the mysterious $C_{10}H_{15}N$ represents the chemical formula of methamphetamine. But how about the meaning of the number 149.24? Take a guess: a) the molecular weight of methamphetamine; b) the water temperature at Bogdan's car wash; or c) Bryan Cranston's weight at the end of the show. Yes, you guessed right.

There is more in this book than science. Dr. Nelson describes a few elements of her experience as a science consultant for *Breaking Bad*, from getting in touch with the production to advising the writers. She and I share an admiration for Vince Gilligan, the creator of what is arguably one of the best shows in television history. In my case, the admiration started on a fateful Albuquerque afternoon when, after a serendipitous yet lengthy discussion, Vince decided to give my eyebrows a chance. And what a ride it was! I learned a lot and made good friends. Even today, when Bogdan the car wash owner is walking the streets of Chicago, people stop him to chat or take pictures. Some know that the bushy-eyebrows guy who gave virtual life to the character is a scientist by profession, and some don't and don't care.

Now here is a word of caution: before breaking bad yourself, carefully read this book to get your red and white phosphorus right. Or send me a message; after all, I have a PhD in chemistry.

And "IT'S CHEMISTRY, BITCH!"

Dr. Marius S*Ta*N

Acknowledgments

The Science of "Breaking Bad" began its journey as a pop culture article on Collider.com in the fall of 2013 and is now the very book that you hold in your hands. I thank Collider's editorial team for carving out a space for me to publish that original science explainer, which has grown in size, scope, breadth, and depth over the years on the path to becoming this book. I have MIT Press acquisitions editor Jermey Matthews to thank for reaching out with the opportunity to make this book a reality. It goes without saying that the show *Breaking Bad* itself wouldn't have been nearly as scientifically sound as it was without the consulting work of science advisor Dr. Donna J. Nelson; this book may not have happened at all without her involvement as a coauthor, so she has my thanks, as well.

My deepest gratitude goes to Allison Keene, whose love, patience, and support made it possible for me to finish this book. And many thanks go out to copy editor Julia Collins and the editorial team of the MIT Press, including acquisitions assistant Gabriela Bueno Gibbs, assistant editor Virginia Crossman, design manager Yasuyo Iguchi, and executive publicist and communications manager Jessica Pellien, all of whom lent their expertise to make *The Science of "Breaking Bad"* the best book it could possibly be.

—Dave Trumbore

I'd like to thank Vince Gilligan and the other producers along with the writers, cast, and crew of *Breaking Bad* for making my experience as science advisor so delightful and interesting. I also appreciate David Trumbore's work on this book and his working with me. In addition, the patience and excellence of the entire MIT Press staff were of great benefit.

—Dr. Donna J. Nelson

1 Meet Walter White

Figure 1.1
Dr. Donna J. Nelson at the *Breaking Bad* office in Burbank. Image courtesy of Chris Brammer.

From Dr. Donna J. Nelson:

My first exposure to *Breaking Bad* came from an article in *Chemical & Engineering News* (March 3, 2008, pages 32–33), the weekly magazine of the American Chemical Society ("'Breaking Bad': Novel TV Show Features Chemist Making

Crystal Meth"). In this article, Vince Gilligan, the producer of *Breaking Bad,* was interviewed during Season 1. My eye was caught by the article's photo of the lead character Walter White standing in the desert, wearing only underwear, shoes and socks, and a lab apron—not lab apparel commonly seen. Upon reading the article closely, I noted a statement by Vince: "We welcome constructive comments from a chemically inclined audience." This was an opportunity we had anticipated for a long time. Interested parties from the scientific community to the U.S. Congress had discussed the goal of influencing a prime-time television show based on science, as a way to increase the public interest in science. We couldn't see how to have such influence, but now, it was being suggested by a Hollywood producer!

But I hesitated because of the show's connection with illicit drugs. I worried that connection could be damaging to my academic reputation; I had carefully ensured it was well known that I had no connection with anything illegal, in order to serve as a proper role model to my students. I initially dismissed the producer's invitation to weigh in.

Then, losing this unique opportunity gnawed at me; it probably would never come again. I decided to test my assumption that *Breaking Bad* might present illicit drugs as positive. There were only five episodes of the show existing at that time, so I carefully watched them all. I saw that each portrayed illegal activity that was eventually punished; the show was ultimately moral. Then I felt more comfortable with the show—sufficiently comfortable to offer to help.

However, I still had more worries. I knew that sometimes people made statements for the sake of appearance, with no intention of following through. Also, I had no way to contact Vince.

So, I contacted the reporter who authored the *C&EN* story, asking if she thought Vince was serious. She didn't know; she had not thought of the statement as an opportunity for us to influence Hollywood and its messages to the public. However, I knew she had his contact information. I asked if she was willing to communicate to him that I volunteered to help with the science. The *C&EN* article stated that *Breaking Bad* had no funds to hire a science advisor, so I knew I wouldn't be paid, but it was still a fabulous opportunity to serve the scientific community.

The reporter said she would contact them, and about one week later, I received a phone call in my office. It was a *Breaking Bad* representative, asking if it was true that I would help them. I said, "Yes." Then she asked if I ever made it out to Burbank, California. I knew they had no funds to support my travel, so how I answered would have a large impact on whether the collaboration moved forward. Therefore, even though I hadn't been to Burbank, I replied, "Oh, all the time," and she said, "Great, the next time you are in town, come by to meet us." And with that, our collaboration started.

My name is Walter Hartwell White. I live at 308 Negra Arroyo Lane, Albuquerque, New Mexico. 87104. To all law enforcement entities, this is not an admission of guilt. I am speaking to my family now.—Walter White, *Breaking Bad,* Season 1, Episode 1: "Pilot"

On January 20, 2008, AMC introduced its newest TV drama series in the following manner:

A pair of khaki pants, still belted, floats through the air and lands on a dusty road somewhere in the desert wilderness of the American Southwest. The pants are soon run over by a recreational vehicle being driven recklessly by a middle-aged white man wearing only his wedding ring, underwear, and a full-face respirator. His passenger, also wearing a face mask, is passed out in the front seat. Inside the RV's main compartment, bodies of two men slide back and forth across the chemical-slicked floor as laboratory glassware shatters around them. The RV crashes, the driver stumbles out, and, once he composes himself, he records a video message for his family in the unfortunate event that he doesn't make it home alive. The man, now known to the audience as Walter White, then aims a gun down the road as sirens approach. Cut to the show title card.

The million or so viewers who tuned in to watch this episode at the time had questions: Who exactly was Walter White? Why was the guy from *Malcolm in the Middle* standing in the desert in his underwear? What, exactly, were we watching?

This was *Breaking Bad,* the brainchild of Vince Gilligan, then a two-time Emmy nominee for his role as producer/writer/director on the smash-hit sci-fi series *The X-Files,* who also had three feature film scripts under his belt. His vision for *Breaking Bad* was built around a narrative little seen on network or even cable TV up until that point. Rather than create a likeable protagonist who could stay morally and philosophically static year after year and season after season, Gilligan wanted to "take Mr. Chips and turn him into Scarface," to paraphrase his go-to description. He envisioned a man who would turn away from his otherwise normal life, a man who would "break bad" and commit to a life of crime, specifically, the illicit manufacturing of methamphetamine.

That man, the same man standing in his underwear in the pilot episode, was Bryan Cranston. He would soon leave his earlier role of immature father-figure Hal Wilkerson from *Malcolm in the Middle* in the New Mexico dust in order to take on the role of Walter White, a mild-mannered high school chemistry teacher who sets out on the path to becoming the notorious drug kingpin known only as Heisenberg. But in the real world, neither Gilligan nor Cranston were scientists, much less chemists, and neither were any of the show's writers. So how on Earth did they expect to pull off such an ambitious story in a convincing way?

During the inaugural episode of the excellent *Breaking Bad Insider Podcast,* hosted by the show's editor Kelley Dixon and creator Vince Gilligan, Gilligan himself stressed the importance of getting the science right:

> I'm no chemist. I never took chemistry in high school. I kinda wish that I had, except that I don't think I'd like it as much now if I did. There's something beautiful about math and science because there are right or wrong answers, and then the rest of life, as we all know, everything's just a big gray area.
>
> None of my writers, actually, is a chemist. None of them know anything more about chemistry than I do. We have Kate Powers, we have a wonderful researcher in Gennifer Hutchison. We have Dr. Donna [Nelson] at the University of Oklahoma who's helped us out a great deal. ... We had a DEA chemist who happened to be on a break from work who came and visited us on set. ... They didn't show us how to make meth, but they told us when we were technically wrong because we wanted to get this right.[1]

Gilligan's commitment to bringing real-world science to *Breaking Bad* was commendable, but it was also a smart creative decision that helped the show to stand out from the pack. As the show's popularity—and its ratings—climbed over the years, it became very apparent that audiences were fascinated by Walt's mastery of science and were curious to know more about his explosive experiments and meth-making machinations. So in addition to the slew of episode recaps and reviews that sprang up across entertainment news websites, magazines, and podcasts, there was also an uptick in the coverage of the show's science. One such fact-checking article was my own 2013 effort for Collider.com titled, "The Science of *Breaking Bad.*" As you might have guessed, that article ultimately led to the writing of this book, which offers a much deeper dive into the show's excellent grasp of scientific concepts.

The only issue with offering a portrayal of actual scientific processes in a show like *Breaking Bad*—which features illicit meth making, malicious use of explosives, and gruesome methods of "evidence" disposal, to name a few—is that ne'er-do-wells might get the idea that they know enough just by watching the show to become their own Heisenberg. While it should go without saying that the laws of the land trump both the possibilities of fiction and the practicalities of science, consider this our disclaimer: **none of the information in this book is intended to be used in an illicit, illegal, or ill-advised manner.**

And as Gilligan and the creative team behind *Breaking Bad* had already figured out, simply watching the show probably wouldn't get you very far anyway. As Gilligan put it:

> I don't want people leaving this show knowing how to make meth. We never wanted this to be a how-to about making ricin, or meth, or any of this stuff. It's a story of a transformation from a good guy to a bad guy.[2]

This is all to say that *Breaking Bad* is a **fictional** show with a basis in real-world science that was tightly scripted, controlled, and produced with all necessary legal and safety precautions put into place. Viewers only get to see the sexy side of the production without all the camera trickery and special effects work that goes into it later. *Breaking Bad* is not a how-to guide and neither is *The Science of "Breaking Bad."* This book is intended to give fans of the show a deeper knowledge of the real-world science on display throughout, to clarify just how close the TV science comes to reality, and to reveal the great lengths that the show's creative team went to in order to get it right.

The writers of *Breaking Bad* employed staff researchers like Jenn Carroll, Kate Powers, Gordon Smith, and Gennifer Hutchison to run down the nitty-gritty details of the science they'd be delving into. The writers also spoke to a number of experts, including Drug Enforcement Administration (DEA) agents, DEA senior chemist Victor Bravenec (whose name was given to the thoracic surgeon who removes Walt's lung tumor), radiation oncologist Dean Mastras (who happens to be writer/director George Mastras's brother), spokespersons for Narcotics Anonymous, the Albuquerque Police Department, the officers of the New Mexico State Police and their drug-sniffing dogs, AMC executive Brian Bockrath (who holds a degree in electrical engineering), and a retired freight rail hazmat safety specialist who basically wrote the book on the topic. Even Bogdan Wolynetz, the bushy-browed owner of the A1A Car Wash in the show, is actually a computational physicist and chemist by the name of Marius Stan who is a senior scientist at the Argonne National Laboratory after having worked at Los Alamos National Laboratory.

And, of course, the show's longtime science advisor and consultant—and this book's coauthor—Dr. Donna J. Nelson has been an integral part in establishing *Breaking Bad*'s scientific literacy. As noted in the preface, Nelson came aboard the production as a volunteer after reading a chemistry-focused interview with Gilligan and Cranston in *Chemical & Engineering News* back in 2008[3] during the show's first season. In the article, they put the call out for scientists, especially chemists, to lend their expertise to the show; Dr. Nelson was the only person who followed up.

It's a good thing she did! Dr. Nelson has been instrumental in helping the show's creative team lock down their organic chemistry and methamphetamine syntheses (along with proper precautions taken to keep the show from being a how-to guide), and even crunching numbers for dilution problems on the scale of freight train tanker cars. A professor of chemistry specializing in organic chemistry at the University of Oklahoma, Dr. Nelson is a Guggenheim Fellow and a Fulbright Scholar and was the American Chemical Society President in 2016. I'm thrilled and honored (and honestly relieved) to say that she has fact-checked the science presented here and has graciously shared insider information and personal experiences from the show; her anecdotes can be found throughout this book.

Although *The Science of "Breaking Bad"* is intended for fans of the show who have a curiosity or enthusiasm for science, I don't want to exclude science enthusiasts who might never have seen the show but have a passing curiosity for it. So allow me to introduce you to Walter White, a.k.a. Heisenberg.

The pilot of *Breaking Bad* is still the best possible introduction to the show and its polarizing protagonist, but there are many moments scattered throughout the series that act as both science lessons and clever displays of Walter White's dual nature. Many of these are active, such as Walt's various MacGyver-like contraptions—rigging a DIY battery, preparing homemade explosives, cooking meth—but others are subtler. Take, for example, Walt's high school chemistry lesson on chirality in Season 1, Episode 2: "Cat's in the Bag."

As Walt tells his students, "chiral" is derived from the Greek for "hand" and is, fittingly, a description that's best illustrated by comparing the shape of your right hand versus that of your left hand; they're identical yet opposite and non-superimposable. Basically, they're mirror images of each other. This trait also exists all the way down to the molecular level of matter. Walt, in a very telling bit of dialogue, says, "Although they may look the same, they don't always behave the same." It might not be apparent right away, but that line is an apt description of Walter White and his alter ego, Heisenberg. That is some stellar foreshadowing disguised as a science lesson. *Breaking Bad* does this better than anyone.

But the lesson doesn't stop there. Walt goes on to use a concrete, if tragic, example of chirality: the drug thalidomide, used in the mid-twentieth century to prevent morning sickness in pregnant women. This particular version of the molecule's shape—known as an isomer—is the "right-handed" one and it's perfectly safe; the left-handed isomer, however, can produce severe birth defects if taken. This is precisely what affected some ten thousand children across forty-six countries in the 1950s and 60s. The scandal eventually strengthened regulations in organizations like the FDA, which required proven efficacy and disclosure of side effects for future drugs. It's a stark lesson on the nature of chirality and one that provides the dark narrative undercurrent running throughout the series as a metaphor for the harmless Walter White and the deadly Heisenberg. They may look the same, but they certainly don't behave the same way.

You might be wondering why Walt chose that name as an alias, even if you've already seen the series in its entirety. While "Heisenberg" works perfectly well as a pseudonym, it works doubly well when you understand that it's a reference to Werner Heisenberg, the twentieth-century German theoretical physicist and pioneer of quantum mechanics, for which he won the Nobel Prize in Physics in 1932 at the age of only thirty-one. His contributions to theories related to the atomic nucleus and subatomic particles, nuclear reactors, and nuclear weapons are certainly remarkable and worthy of their own separate study, but Heisenberg is best remembered for one thing: the uncertainty principle.

This characteristic of quantum mechanics basically means that the closer you get to measuring the true position of a particle, the less sure you are about its momentum, and vice versa. In science, this is tricky enough, but in the world of *Breaking Bad*, it's downright dangerous. Walter White chooses this specific alias early on in part because it sounds badass (especially when paired with dark sunglasses and a black pork pie hat), but even more so because of its connotation: You might think you have Heisenberg pinned down, but that's when he'll blindside you with something unexpected ... and deadly.

More to the point, the show's writers chose the name Heisenberg for both its dramatic and scientific value (and on the more morbid side of things, because the scientist himself also battled cancer before passing away in 1976). The cast and crew have taken great pains to walk the walk and talk the talk of real-world chemists, physicists, and engineers. It's one of the many impressive features of *Breaking Bad* and serves as the inspiration for this very book.

In the following chapters, I'll lead you through an exhaustive list of scientific moments found throughout the five seasons of *Breaking Bad* to explain them in both simple terms (see the 101 section in each chapter) and using a little more advanced terminology (see the Advanced sections, as well as the glossary of terms at the end of the book). We'll be covering chemistry, biology, and physics as portrayed in *Breaking Bad,* and more specifically we'll get into the subdivisions of each area of study, like meth making, toxicology, and electromagnetism.

This book is set up in such a way that you can read cover to cover, start to finish, or you can jump to the subsection that interests you most (though the three successive chemistry sections do build upon each other). Throughout the book, you'll also find bits of trivia under the "Inside *Breaking Bad*" heading, which aims to show the Hollywood side of the show's attempts to portray real-world science. The incredible talent assembled behind the scenes of *Breaking Bad* went to great lengths to get the science right and greater lengths still to portray it in a safe but realistic manner on camera. I'd be remiss if I didn't call attention to those efforts.

Below, you'll also find the first of fifteen bits of bonus content I like to call "Side Reactions"—shortened to "Side RxN." These are worthy scientific observations seen throughout *Breaking Bad* that didn't quite fit in an overall chapter. Think of them as bite-sized portions of the show's science that you can consume in a single sitting. Their topics range from the chemical composition of the human body, to "zombie" computers, to what happens when balloons tangle with power lines. In other words, it's the extra-fun stuff. But don't take my word for it: check out the first one for yourself!

Side RxN #1: Crystallography and Synchrotrons

In another life, before cancer, meth, and the assumption of the Heisenberg identity, Walter White was a brilliant cofounder of Gray Matter Technologies along with fellow grad school colleague Elliot Schwartz. ("Schwartz" is German for "black"; Black plus White = Gray.) Walt sold his share for $5,000, only to watch the company go on to be valued at $2.16 billion (with a "B"). That fact irritates Walt from the show's premiere to its finale, growing from a regret, to a grudge, to a perceived injustice. The stark difference in how their lives turned out is laid bare when the Whites pay a visit to the Schwartzes' swanky mansion for a party in the appropriately titled episode "Gray Matter."

There, a former grad school research partner named Farley introduces Walt to some other party guests as "a master of crystallography" back in their Caltech days. Crystallographers seek to determine the arrangement of atoms in a crystalline solid, like a protein, through the use of x-rays along with neutron diffraction and electron diffraction. Think of the process as being like shining a flashlight beam onto an object in the dark to figure out exactly what it is; this is the same idea, just on a molecular scale.

When the Caltech researchers were presented with a "protein problem," Farley recalled, Walt had a one-word solution: "synchrotrons." As Walt explained to the group in the episode, "They generate purer and more complete patterns than x-ray beams. Data collection takes a fraction of the time." A synchrotron is a type of particle accelerator in which a beam of charged particles travels through a closed loop where it is steered and focused by electromagnets; this magnetic field is "synched" with the beam as it accelerates. (CERN's Large Hadron Collider is the world's largest synchrotron-type accelerator.) Synchrotron radiation is emitted during the acceleration process and is highly stable, of high intensity, and polarized—or focused—allowing for data collection that takes less time than using weaker x-ray sources. Walt got this one right. We'd expect no less from the man who contributed research toward a Nobel Prize in Chemistry, though he wasn't awarded the prize himself. This perceived injustice not only gave Walt another reason to hold a grudge against his former partners but also nudged him down the path toward becoming Heisenberg.

Chemistry I

It should come as no surprise that there are going to be a lot of chemistry discussions in this book. Like, *a lot.* It's the basis of Walt's career, whether we're talking about his days in Gray Matter Technologies, his time as a high school chemistry teacher, or his drug empire business. It's also the fundamental science that runs parallel with the show's narrative: While molecules transform, elements decay, atoms break bonds, and electrons drive change, the same can be said for Walter White; his alter ego Heisenberg slowly takes over to become his default personality, or "ground state." (That's another term for the lowest energy state of an atom. See? We're learning chemistry already!)

If you've never taken a chemistry class in your life, don't fret; we're going to start out slowly and cover the basics. If you're an accomplished chemist with years of lab work and stacks of published papers to your name, I think you'll find something worthwhile in the following chapters simply because of how they tie into *Breaking Bad.*

In the first chemistry section, I'll be revisiting *Breaking Bad*'s famous opening credits sequence, which uses the highly recognizable periodic table of elements as a clever way to introduce the show's title, cast, and crew. (There's a lot more going on here than just elemental symbols and a now-iconic color scheme.) Next, we'll play with fire. Not literally, of course—we'll leave that to the professionals. But Walt's fiery demonstration of colored flames in his chemistry class is a colorful way of explaining how to excite electrons and get them to put on a show. The third and final subsection gets a little more dangerous; it's where we start to "break bad" for real. I'll take a look at Walt's deadly phosphine gas concoction to see just how vile the vapors really are.

II Chemical Credits

From Dr. Donna J. Nelson:

I enjoyed my set visits to Albuquerque; it is interesting that the show was originally supposed to take place in Riverside, California, but it was moved to Albuquerque when New Mexico gave a 25 percent tax rebate to film production. During my first set visit in Albuquerque, Breaking Bad's producer Vince Gilligan asked a question about using chemical symbols in the opening credits, which was very important to the show.

He mentioned that they wanted to use the symbol D in one of the names listed in the credits, but he had been told that there was no element that had that symbol. He asked me if that was correct. I replied that it was, but that if he wished to be creative, he could use the symbol for deuterium, an isotope of hydrogen. I also mentioned at that time that there is another isotope of hydrogen, which is named tritium, which has the symbol T, so he could use that in the credits also. He said he would keep this information in mind for future reference. I was to learn that Vince was a science groupie. He had learned a great deal of science on his own, and he truly enjoyed scientific experiments. He would demonstrate to Bryan Cranston, who portrays Walt, and Bryan would then act out the experiments on camera.

When I first saw it presented in the show, the elemental composition of the human body seemed to me potentially confusing. This is because these data can be expressed in multiple ways, so one must keep in mind exactly what these data pertain to. Data on elemental composition of the human cell are also available, but the data in Episode 2 are for the human body, so they are not the same. Also, the data in Episode 2 are averages by mole or by the number of atoms present for each element. They are not percentages by mass, which are frequently used and which will be vastly different from those in Episode 2, because the masses of the elements differ considerably from each other.

101

Chemistry is, well technically, chemistry is the study of matter. But I prefer to see it as the study of change.—Walter White, Season 1, Episode 1: "Pilot"

One of the most iconic aspects of *Breaking Bad* is the show's opening credits sequence. The narrative is steeped in chemistry and the production team wastes no time in reminding audiences of that fact each and every time an episode starts up. If you've found yourself wondering just what those bold, stylized letters in the names of the cast and crew are all about, or what the assortment of numbers in the green boxes mean, that's as good a place to start our discussion of the science of *Breaking Bad* as any.

There's a lot more going on in the original opening credits than you might think. The word "meth" appears and then soon disappears against the smoky green backdrop; this one should be pretty self-explanatory considering the entire plot is built around the illegal drug. What's less obvious, however, is the meaning of the number 149.24 or the alphanumeric soup that is $C_{10}H_{15}N$.

Fun Fact: These both represent meth as well, just in different ways. The number is the molecular weight of methamphetamine; the string of letters and numbers is meth's chemical formula, revealing that it's composed of the relatively common elements of carbon, hydrogen, and nitrogen.

Some less-familiar elements arrive elsewhere in the credits sequence, most notably, of course, Br and Ba, which are the first letters in the show's title and act as an abbreviation for *Breaking Bad*. However, in the chemical world, they also symbolize the elements bromine and barium. Other elemental symbols appear in the credits, like C for carbon, V for vanadium, and Cr for chromium, just to name a few. They're styled after the familiar symbols found on the periodic table of elements, but that's not where the similarities end. A sharp-eyed viewer will spot a number of, well, *numbers* surrounding the symbols of Br and Ba. Without going into too much detail just yet, these numbers tell us a lot about the elements themselves. The atomic number, relative atomic mass, and oxidation states are all more or less reproduced faithfully in *Breaking Bad*'s opening credits; the only minor error occurred early on in the elements' electron configuration, which was corrected in later episodes. That should give you a general idea of what *Breaking Bad*'s crazy credits are all about, but for a deeper dive into what they represent, read on.

Advanced

The presentation of the periodic table of elements communicates a wealth of information in a simplified format, but it helps to understand the nature of elements and atoms to begin with. First published by Russian chemist Dmitri Mendeleev in 1869, the periodic table has evolved over the last 150 years to include 118 elements that have either been observed in nature or have been synthesized in labs or nuclear reactors. Research on synthesizing elements of even higher atomic numbers is currently underway.[1]

So what exactly is an "atomic number"? It's the number of protons in the nucleus of an atom of a given element. That's why hydrogen, with its single proton, has an atomic number of 1, while the recently synthesized oganesson and its 118 protons has an atomic number of 118, the lowest and the highest of these numbers respectively. Since protons are positively charged subatomic particles, the atomic number also represents the charge of a given nucleus, which contains only protons and neutrally charged neutrons. In an uncharged atom, the atomic number also matches the number of electrons in the electron cloud; since these subatomic particles are negatively charged, equal numbers of protons and electrons cancel each other out when it comes to charge.

This is a good time to talk briefly about electrons and how they operate within an atom and in relation to other atoms. These concepts are at the heart of all chemical reactions and interactions, serving as the foundation to all the chemistry talk that is to come in future chapters. I like to think of electrons as the currency of the atomic world. If an atom is a businesswoman whose only goal in life is to get the best possible bang for her buck, then electrons are the money she spends and receives to achieve this; the electron cloud, which surrounds the atom's nucleus, would be her bank account. A stable bank account is of utmost concern, though a little financial flexibility is possible.

A businesswoman with a depleted account is eager to bring in some more cash, while someone else with a lot of money to spend is more willing to give some of it away. Atoms with either too few or too many electrons in their cloud act in a similar fashion, tending to take up or give up electrons, respectively. The electron cloud of a given atom is made up of energy levels, or shells, with the lowest energy shell existing closest to the nucleus. These shells are further broken down into atomic orbitals, which are regions of space around the nucleus where an electron or electron pair is 90 percent likely to be found, as defined by a mathematic equation. It might help to think about these shells and orbitals as subaccounts and partitions within a larger bank account.

When it comes to an atom's stability, whether we're talking about hydrogen and its solitary electron or oganesson and its 118 electrons, it's the atomic number that represents each elemental atom's most stable isotope, or variant. In extreme cases, atoms will behave violently in order to give up or take on electrons as quickly as possible, which is something to keep in mind for our upcoming discussion of explosives ...

But *Breaking Bad* is most concerned with atomic numbers 35 and 56, representing bromine and barium, respectively. The specific elements don't have much to do with the plot at all, they're just featured—cleverly so—since they match up with the show's title. However, the production team put a lot of effort into faithfully reproducing the periodic table of elements for the title sequence well before they ever knew the award-winning show would go on to enjoy sixty-two episodes. They put in so much effort, in fact, that the elements on display not only include the correct atomic number, but also the proper relative atomic mass and oxidation states. They even attempted to show off the elements' electron configuration, with some slight errors; we'll get to that in a moment.

An elemental atom's relative atomic mass seems like a pretty straightforward measurement, but like most values on the periodic table, it's a simplified representation of something more complex. It's the ratio of the average mass of atoms of an element in a given sample to one unified atomic mass unit, symbol u, which is defined as one twelfth of the mass of a carbon-12 atom in its ground state.[2] Since most of an atom's mass exists in its protons and neutrons, and since the same element can differ in its number of neutrons (i.e., isotopes), an atom of a given element can have differing masses, which is why an average of these masses is used to calculate the relative atomic mass. For bromine, this value is 79.904, and for barium, it's 137.327, both of which can be seen in the show's credits. It's a pretty obscure measurement for anyone outside of chemistry research and industry but a cool addition from *Breaking Bad*'s creative team nonetheless.

Another super nerdy inclusion from the *Breaking Bad* title credits is the oxidation state of elements. This value, also called the oxidation number, indicates the degree of oxidation—or loss of electrons—of a given atom; this number is very useful when balancing chemical reactions. The value is governed by a set of rules, but conceptually it illustrates the charge an atom might carry under a given set of conditions. Elements can have a wide range of oxidation numbers representing both positive and negative charges, depending on whether the atom is taking on electrons or giving them up, but tables often list the most commonly occurring states. *Breaking Bad* opts for this space-saving method by only listing +1, −1, and +5 for bromine, and +2 for barium.

The last bit of fun the *Breaking Bad* team had with the periodic table of elements drew the critical eye of chemists and chemistry enthusiasts. The production team opted to include each element's electron configuration as part of the credits as well. An electron configuration is a string of numbers and letters representing the distribution of electrons in atomic orbitals, or if you prefer the businesswoman analogy, how much money is in each uniquely named subaccount or partition. While the show's representation of the periodic table in the background of the credits accurately lists elemental electron configurations (though in a simplified numeric format that only states the number of electrons per orbital without naming said orbitals), an error occurred early in the series when individual elements were highlighted; this wasn't corrected until well into Season 5. Still, not too bad considering how much thought went into something as fleeting and easily overlooked as a show's opening credits!

Inside *Breaking Bad*: Despite the attention to detail by the production team, mistakes happen. The fictional show doesn't always reflect real life, either by accident or by practical necessity. Even the credits of *Breaking Bad* fell victim to human error.

One example of this occurred when *Breaking Bad* tried to use the abbreviation Ch in Michael Slovis's credit as the show's director of photography. However, no such elemental abbreviation exists. Oops! It was later changed to C for carbon.

Another example was found in the aforementioned electron configurations. The production team nailed the distribution of electrons per shell for bromine (2, 8, 18, 7), but erroneously copied this same sequence for both barium and chromium (Cr); the latter element was seen in the credit "[Cr]eated by Vince Gilligan." The proper electron configurations of barium (2-8-18-18-8-2) and chromium (2-8-13-1) were used when the second half of Season 5 aired, delighting many concerned chemists out there in TV land.

Side RxN #2: Chemical Composition of the Human Body

Gray Matter Technologies actually had a third member in addition to Walter White and Elliot Schwartz: Gretchen Schwartz, Walt's former lab assistant and love interest who ended up marrying Elliot after Walt suddenly abandoned her. During their halcyon days together, however, Walt and Gretchen had a science-meets-philosophy discussion about the elemental composition of the human body in Season 1, Episode 3: "... and the Bag's in the River." Despite the flirtatious nature of this conversation, brilliant editing pairs it with Walt and his meth business partner Jesse Pinkman's contemporary clean-up of the mostly dissolved body of Jesse's previous partner, Emilio Koyama, bit by bloody bit.

Back when their romance was still on, Walt and Gretchen wound up with the following elemental percentages by mole, or atomic percent, written on a chalkboard:

Hydrogen (63 percent), oxygen (26 percent), carbon (9 percent), nitrogen (1.25 percent), calcium (0.25 percent), phosphorus (0.19 percent), sodium (0.04 percent), and iron (0.00004 percent), plus the chalkboard notations of chlorine (0.2 percent), sulfur (0.050002 percent), and magnesium (0.00404 percent), which Walt says adds up to 99.888042 percent, meaning 0.111958 percent is unaccounted for. Walt also says, "It seems like something's missing, doesn't it? There's got to be more to a human being than that."

While this philosophical pondering could be intended to encourage the audience to question Walt's own soul, there's a bit more to the body than what was mentioned in the episode. (The commonly held belief that the human soul weighs about 21g [grams] was based on an attempt to determine the value scientifically by Dr. Duncan MacDougall in 1907, but the consensus is that this is more of a curiosity than a fact. I kindly point you to the nearest Google machine if you'd like to know more.)

The previously noted percentages by mole are correct, except that chlorine is present at 0.025 percent, sulfur at 0.06 percent, and magnesium at 0.013 percent, according to available literature. Also absent from this list are potassium at 0.06 percent and iodine at—stick with me—0.000002 percent. The show's math is wrong for the numbers on the chalkboard, but their end value of 99.888042 percent checks out when factoring in the correct numbers. That still leaves us with more than a tenth of a percent unaccounted for.

As Gretchen herself says, "That only leaves you with the trace elements, down where the magic happens." Along with iron and iodine, trace elements include less than 0.01 percent each of boron, cadmium, chromium, cobalt, copper, fluorine, manganese, molybdenum, selenium, silicon, tin, vanadium, and zinc. Taken together, they add up to less than the percentage of magnesium in the body, though there's certainly a little wiggle room for variation among people. Amounts of trace elements may be small, but they are mighty (and often essential), like iron's critical role in the oxygen-transporting molecule hemoglobin, or iodine's role in proper thyroid hormone formation.

III Playing with Fire

From Dr. Donna J. Nelson:

Although Breaking Bad was filmed in Albuquerque, the writers and main office were located in Burbank, California. I was invited to visit that office during my initial phone conversation with the Breaking Bad representative.

By coincidence, Chancellor Marye Anne Fox, at the University of California, San Diego, also named me their Chancellor's Diversity Scholar. As an untenured assistant professor, she had been on my graduate committee at University of Texas at Austin, while I earned my PhD there, and we have been good friends ever since.

During the first week at UC San Diego, I rented a car, drove to Burbank, and kept an appointment with Vince Gilligan at 11 am. I took with me my son Chris, who is a chemical engineer. During the flight to San Diego and the drive to Burbank, I constantly told my son not to get his hopes up about the scheduled meeting, because there were no guarantees that we would actually meet Vince or anyone else truly influential to the show. I was avoiding disappointment for myself as much as for him.

Upon entering, we found the Breaking Bad offices were rather small, and the interior was spartan, decorated solely with Breaking Bad posters around the hallway leading to the writers' rooms.

I introduced myself to the receptionist, offered my business card, and asked for one belonging to anyone in the offices. She laughed, saying, "No one uses business cards in Hollywood." Then she phoned, telling Vince I had arrived.

In less than a minute, Vince rushed out of an adjoining room, flashing a smile and holding out his hand to shake mine. He showed us into the adjoining room, which he and his writers used to discuss and plan the plot for each episode. There were columns of three-by-five-inch index cards tacked on a large bulletin board and the walls, each bearing a list of short phrases. It appeared that each column represented an episode. He had me sit at the head of a large table, with my son on one side and him on the other. He said the writers wanted to meet us too, and momentarily, they entered and populated all the seats around the table.

For an hour, I answered their questions. I didn't realize at first, but their questions were about character development. It was the end of Season 1, and they were still developing Walt's personality and those of the characters around him. Examples of questions were "What kind of person becomes a scientist?" "How does a chemist talk to people?" "How does a professor talk to students?" "Does a professor talk to students in the lab differently than outside the lab?"

As lunchtime approached, I assumed our meeting would end, so I was delightfully surprised when they invited us to lunch with them. We drove a short distance to Gordon Biersch Brewing at 145 S. San Fernando Blvd., where brainstorming continued for another hour.

One specific question was "What would make a person who obtained his PhD from a terrific university become a high school teacher, while another student collaborating on his research project went on to win a Nobel Prize?" I asked if a woman was involved in the research group, and they said yes. I told them to have his coworker take away his girlfriend, and it would crush him; he would never recover and drop out. Later, I would recognize the significance of this question.

At the end of lunch, they asked if I'd be willing to answer additional questions posed by email or by phone. I answered, "Of course!"

101

Electrons, they change their energy levels. Molecules change their bonds. Elements, they combine and change into compounds. That's all of life, right? It's the constant, it's the cycle. It's solution, dissolution. Just over and over and over. It is growth, then decay, then transformation. It is fascinating really.—Walter White, Season 1, Episode 1: "Pilot"

If the opening credits are the appetizer for the amount of science in *Breaking Bad*, then the show's pilot episode acts as the main course. Walter White not only cooks up his first-ever batch of crystal meth and uses his scientific knowledge to take down some ruthless drug dealers in this first hour, he also has the time to teach his classroom full of disinterested students about the transformative power of chemistry. Now one of those activities might sound a little less dramatic than the others, but Walt's lesson on chemical reactions as the basis of all life also serves as the philosophical underpinning for the entire series.

Since Walt knows that his teenage students have other things on their minds than electron orbitals and photon emissions, as evidenced by the couple who'd rather flirt than focus, he attempts to grab their attention through a colorful demonstration using the most ancient of scientific curiosities: fire.

While waxing poetic about the nature of chemistry, Walt lights a Bunsen burner and casually sprays different solutions into the flame, turning it alternately red and green for a short period of time. This visually interesting display of chemistry and physics at work wasn't just a cool Hollywood effect, it's something that's relatively easy to reproduce. But what exactly is going on?

In chapter 2, I talked about the nature of electrons within atoms, which compose elements, which comprise all matter. But like Walt says, these electrons and atoms aren't just sitting still, they're shifting, changing, combining and breaking apart again in a never-ending cycle. But at the same time, energy can neither be created nor destroyed, according to the first law of thermodynamics, also known as the law of conservation of energy. It can, however, change forms or transfer from one place to another.

One of the simplest demonstrations of this atomic activity is the introduction of a solution of metal salts (usually dissolved or suspended in alcohol) to flame. The heat from the fire temporarily excites the electrons in the solution, much like a cup of hot coffee temporarily boosts the energy level of an office worker (an effect that is due more to caffeine—which I'll discuss later—than temperature in this example, but you get the idea). It's inevitable that the caffeine crash will eventually knock the worker back down to normal; so too does the excited electron fall back to its own normal resting state once the energy source, the heat, has been removed. Since energy can neither be created nor destroyed, the extra oomph has to go somewhere. In this case, it escapes in the form of a light-emitting photon. Presto chango, you've got colored fire!

Advanced

There is a *lot* of science packed into this simple demonstration, one that bridges the disciplines of chemistry and physics. Since I introduced electrons and atomic orbitals in chapter 2, let's build on that knowledge to explain what's going on with this fiery demonstration.

In the scene in question, you can see a number of interesting items on the chalkboard behind Walt, if you can tear your eyes away from the pretty flames on display. On one end of the board is a list of electron orbitals in the alphanumeric nomenclature that should be familiar to Chemistry 101 students. On the other end is a series of equations that might look like Greek to you, but once the symbols are defined, the connection between the disciplines of chemistry and physics should be illuminated.

So let's start with the atomic orbitals. As I mentioned in chapter 2, these are the mathematically defined areas around an atom's nucleus where electrons or pairs of electrons are expected to appear. There are different shapes and energy levels of atomic orbitals; they're represented as follows in figure 3.1.

Now this isn't exactly the most intuitive presentation of what atomic orbitals are supposed to look like, but it's a better visual representation than a list of mathematical equations. The rules governing where electrons may fall in these orbitals, however, are pretty straightforward: Due to the Pauli exclusion principle, no two electrons can have the same quantum numbers—which describe the size, shape, and orientation of the orbitals—and only two electrons can be associated with each orbital. Higher-energy shells tend to have more orbitals, so they can hold more electrons overall.

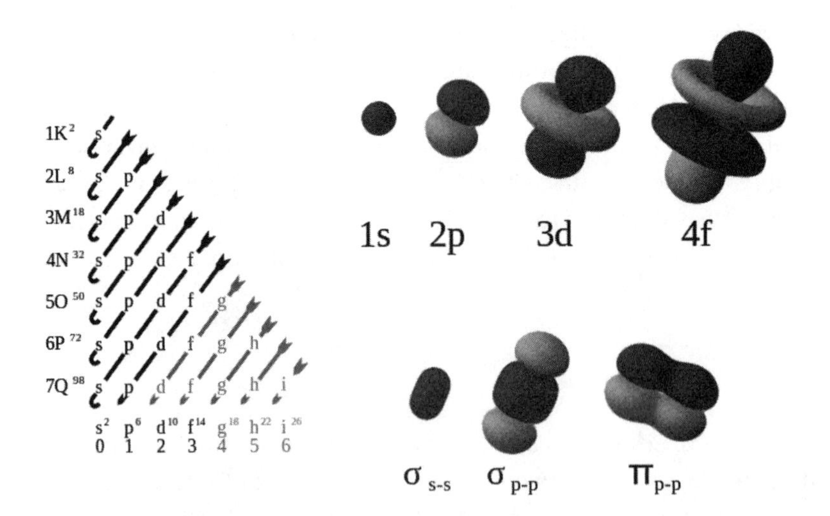

Figure 3.1
Electron orbitals by Patricia.fidi. Public domain, via Wikimedia Commons, https://commons.wikimedia.org.

Importantly, electrons may inhabit any orbital as long as it follows these rules, albeit temporarily, but will ultimately fall to the lowest-energy orbital available. (You can't keep your money in that high-yield savings account if the "Monthly Bills" subaccount hasn't been fully satisfied.) When an electron exists in a higher-energy orbital, perhaps excited by some input of energy, if there's a lower-energy orbital vacancy, the condition becomes unstable. The electron will drop back down to the lower orbital, emitting a photon in the process to account for the loss in energy. This concept is at the heart of Walt's demonstration of colored fire.

Now, not to confuse TV shows, but this jump of an electron from one state to another is known as a quantum leap (or an atomic transition or quantum jump, but those terms are less fun). While the TV series *Quantum Leap* lasted for five seasons, an atomic-level quantum leap can last nanoseconds or less, which explains why Walt's colorful display is such a brief but brilliant demonstration.

But how does this energy difference translate into an emitted photon? That's where the physics equations come in. Here are the equations Walt chose to put on the chalkboard for reference:[1]

$\nu = c/\lambda$
$E = h\nu$
$\lambda = h/mv$

These are nice, concise representations of some pretty complex interactions, so let's start by breaking down the symbols and what they mean:

ν = the frequency of light, measured in hertz (Hz); not a V, but the Greek letter nu

c = the speed of light, a constant; 299,792,458 m/s (meters per second)

λ = the wavelength of light, measured in nanometers (nm)

E = the energy of a photon, a particle that transmits light

h = Planck constant, a value linking a photon's energy and its light's frequency; 6.62606×10–34 J-s (joule-seconds)

m = mass

v = velocity

That first equation tells us that the frequency of light—the number of electromagnetic waves per unit time—is equal to the constant speed of light divided by its wavelength, the measured distance of each complete wave, crest to crest. The frequency and wavelength values determine light's place on the electromagnetic spectrum, from the high frequency/short wavelength values of ionizing radiation (like gamma rays), to the low frequency/long wavelength values of microwaves and radio waves. Between these extremes lies the visible spectrum of light, which exists between the near-ultraviolet and near-infrared sections.

The second equation introduces photons into the mix, a necessary inclusion for Walt's discussion and demonstration. It tells us that the energy of a photon is equal to the Planck constant times the frequency of the electromagnetic wave. So if you know the value of one variable, you can determine the other, making this a very useful relationship when it comes to either identifying the composition of distant stars using astronomical spectroscopy, or demonstrating basic concepts in a high school science class.

The third equation, one of the de Broglie relations, relates the wavelength of light to its momentum (mass multiplied by velocity), giving physicists another tool for determining spectra values.

But what does all of this have to do with rainbow-colored fire? Well, one thing Walt doesn't tell his students is what exactly is in the bottles of liquid he prepared for his demonstration. Walt's seemingly magical concoctions are nothing more than various metal salts dissolved or suspended in an alcohol solution, making for a quick and easy classroom chemistry lesson. (Proper research should be performed on the chemicals ahead of time to determine all potential hazards, and precautions should be taken into consideration and put into effect. I'll also assume Walt had a fire extinguisher at the ready even though we never see one on the screen.) Here's a look at the spectrum of colors available and the related metal salts that could have been used for Walt's demonstration:[2]

medium red = lithium chloride, lithium carbonate

strong red = strontium chloride, strontium carbonate

orange = calcium chloride

yellow = sodium chloride, sodium carbonate, sodium nitrate

yellowish green = sodium borate (a.k.a. borax)

green = copper sulfate, barium chloride

blue = copper chloride

violet = potassium nitrate, rubidium nitrate

purple = potassium chloride

gold = charcoal, iron, lampblack (a.k.a. soot)

white = magnesium sulfate; titanium, aluminum, or beryllium powders

Walt restricts his demonstration to producing green and red flames, probably from solutions of copper and lithium, respectively. Clearly the spectrum contains many more colors than just those two variants. What's fascinating is that, despite the different chemicals and the different colors of flames they produce, the basic concept is the same for all of them: The electrons in the atoms of the compounds dissolved in the alcohol gain energy from being introduced to the fire, causing them to get excited and leap into a higher available atomic orbital. In less than the blink of an eye, this unstable condition forces the electrons to settle back down into their ground state, but since that excess energy has to go somewhere, a photon is emitted. This process is known as spontaneous emission. The energy of this photon, as it relates to frequency and wavelength of the electromagnetic wave in question, ultimately determines its place on the spectrum and, if it's in the visible spectrum, the flame's color.

Interestingly, since the quantum states in an atomic orbital model have discrete values, the differences in energy between these states also have discrete values. So when an electron drops from one orbital to another, the energy value of the photon it emits must exactly match that difference; no more, no less. This phenomenon can be put to use in a process called emission spectroscopy, which determines the elemental composition of an unknown sample by examining the wavelengths of photons emitted as an excited electron transitions back to a lower energy state.[3] Science!

Fun Fact: There's a simple alternative to this lesson that shows off the concept in an even more explosive fashion: fireworks! The next time you see a colorful celebratory display bursting in the night sky, see if you can remember which chemical compounds match up with each color.

While spontaneous emission of light occurs when a substance is heated, there are other forms of nonheated emission known broadly as luminescence. Depending on the manner in which electrons are excited, there are more specific names like electroluminescence (produced electrically), chemoluminescence (produced chemically, of which bioluminescence is a specific type of emission), and photoluminescence, which is light emission after absorption of photons from electromagnetic radiation. Fluorescence and phosphorescence are types of photoluminescence: Phosphorescence is similar to fluorescence except that, rather than instantly emitting photons after absorbing radiation as fluorescent items do, it occurs when excited electrons tumble down to a less-excited, metastable state and slowly emit energy as light even after the exciting radiation is removed. It's basically the difference between fluorescent, UV-activated blacklight posters and phosphorescent, glow-in-the-dark items.

Lasers, ever the scientific curiosity, start by spontaneous emission but are kept going through an even more complicated process known as *stimulated* emission; laser is actually an acronym for Light Amplification by Stimulated Emission of Radiation. *Breaking Bad* never ventured too far into lasers, but they were used to make a pretty powerful point in the series finale.

IV It's a (Phosphine) Gas

From Dr. Donna J. Nelson:

I felt that my helping *Breaking Bad* get the science right was a community service. I believed that my efforts would help me to learn more about how Hollywood functioned, help the producers, directors, and actors learn more about science and scientists, and help the public learn correct science by ensuring that scientific material in the show was correct.

Early on, someone told me that there was a rumor in Hollywood that it was impossible to have both a science advisor and a hit show. I realized then that if I simply aligned my goals with those of the producers, directors, and actors, as I was helping them, that would be sufficient to make a success of our collaboration. I understood immediately that I couldn't let the goal of scientific accuracy impose limitations upon their creative freedom. After all, the show was intended to be a drama—fiction—not a science documentary.

A number of possible limitations crossed my mind. First, the article by which I learned about *Breaking Bad* had appeared in *Chemical & Engineering News*, the magazine of the American Chemical Society (ACS), which went out to all ACS members—about 167,000 at that time. This was obviously going to be a popular television show, and so many people would like to be involved with a television show, I worried that the producers and directors would be swamped with volunteers. They might find that they didn't need me. About a year later during a set visit, I said to Vince Gilligan, "I don't want to talk my way out of a terrific opportunity, but I live in Oklahoma, which is quite a distance from Burbank and Albuquerque. There are so many excellent schools close to you, why didn't you get someone from one of those schools to help you?" Vince replied, "We tried, but no one would help us. We requested a tour of a chemistry lab, but we couldn't get that." Then I asked, "That magazine that featured your interview went out to about 167,000 ACS members; how many volunteers did you get as a result of that story?" Vince looked squarely at me and answered, "One."

Second, I worried that the *Breaking Bad* producers, directors, and writers might decide that they didn't really need me; after all, Walter White was portraying the

Figure 4.1
Dr. Donna J. Nelson's hand-drawn structures for alkenes, replicated in the classroom scene on *Breaking Bad*. Image courtesy of Kate Powers/AMC.

part of a high-school organic chemistry teacher. This was a level of chemistry that is not difficult. The writers were obviously experts at writing, all had been pulled from previous hit shows. Vince had said they had already been pulling science content from the web, so they knew how to do this. I figured they could probably manage to continue without me; they might find they didn't need me.

One of the first requests I got from the writers was about alkene nomenclature. In the scene, Walt was to give a class lecture about this, and they gave me a few script pages to proof. I thought, "Well, this is simple." Then I read the pages and changed my mind, "Oh, no, they do need me." The text on the pages discussed and intermingled alkenes, alkanes, and alkynes—three different classes of simple organic molecules, often with similar names. I applied my typical editing techniques and returned the pages. Then they asked if I could supply any drawings for the blackboard. I gave them structures of a few alkenes, which they reproduced on the board for the scene. My drawn structures can be seen in a photo of the blackboard (see figure 4.1).

Months later, as I watched the episode on the TV in my living room, I was filled with amazement that I had influenced what had just appeared on the screen. Most of all, I felt satisfaction knowing that organic chemistry students (high school or undergraduate) who were watching like me would be reinforced rather than confused by the content. This had been my goal, and I knew then that I would be happy with the arrangement.

101

Red phosphorus in the presence of moisture and accelerated by heat yields phosphorus hydride. Phosphine gas. One good whiff, and ...—Walter White, Season 1, Episode 1: "Pilot"

So now that we have an elemental understanding of the periodic table and have recovered from our excitement over electrons, it's time to get into *Breaking Bad*'s more specific uses of science to save Walt's bacon. Fair warning, it's about to get a little gruesome!

While *Breaking Bad*'s pilot episode spends a good amount of time laying the scientific groundwork to establish Walt's mild-mannered character as a gifted chemist whose contributions led to a Nobel Prize, the most memorable moments come about when Walt puts this knowledge to not-so-good use by synthesizing illegal methamphetamine ... and by gassing a pair of rival drug dealers. I'll get to the science of meth making in a later chapter, but for now, I'll be talking about the dangers of phosphine gas.

For context, in the scene in question, Walt is forced to cook meth by rival drug dealers Emilio Koyama and Krazy-8. He's held at gunpoint and is wearing nothing but a pair of boots, his glasses, an apron, and tighty-whities; he isn't exactly prepared for a fight. Walt does some quick thinking and uses his scientific knowledge of chemistry to cook up a fast-acting poison gas with the chemicals he has on hand. Tossing a jar full of red phosphorus into a pan of hot water produces a towering flame and a noxious cloud of gas that temporarily incapacitates Walt's assailants. That's certainly a quick-thinking way out of a tight spot, but is the chemistry sound?

Phosphorus can indeed react with water to produce the colorless, flammable, and toxic gas known as phosphine, though there are a few more requirements for this reaction, like heat and a hydroxide; it's stable in water otherwise. Phosphine gas is, however, known to be a dangerous byproduct in illegal meth production, so that checks out as well. But while the basics are accounted for here, we find that the devil is in the details.

Advanced

But before I get into those details, I think it's worth learning about the effects that phosphine gas has on the human body. It goes without saying—but bears repeating—that **this book is not an instruction manual or how-to guide for the creation of dangerous and illegal substances.** Perhaps a discussion of phosphine gas's undesirable effects will help to drive that point home.

Phosphine, also known as phosphane and phosphorus trihydride, is a colorless, flammable, toxic gas with the chemical formula of PH_3. According to the chemical's SDS (safety data sheet),[1] the major health hazards associated with exposure to it include: respiratory tract infection, central nervous system depression, and potentially fatal inhalation. Beyond that, phosphine is also an extremely flammable gas that can spontaneously ignite upon exposure to air and can cause a flash fire. Short-term exposure can result in a number of gastrointestinal, respiratory, vision, and cardiac problems, plus liver and kidney damage, convulsions, coma, and possibly death. In liquid form, it's a cryogenic liquid (normal boiling point below –130°F or –90°C) so frostbite can also occur. How's that for hazardous?

Despite that long and deadly warning list, both Emilio and Krazy-8 somehow survive Walt's gas attack ... temporarily, at least. Emilio eventually succumbs to its effects—though how Walt and Jesse get rid of the evidence is a matter I'll discuss later—while Krazy-8 meets his maker in a far more personal manner at the hands of an increasingly villainous Walt. But the lethality of the phosphine gas on the show also falls in line with real-world chemistry since differing levels of exposure to it could result in different health effects. So where does the chemistry in the script go wrong?

It's in the production of phosphine in the real world that *Breaking Bad*'s method comes up short. Phosphine is a pnictogen hydride. This is a fancy name for compounds composed of hydrogen bonded covalently (through sharing of electrons) to atoms of elements from Group 15 of the periodic table. The term "pnictogen" is derived from πνίγειν (pnígein), from the Ancient Greek meaning "to choke," related to the suffocating potential of nitrogen, which you might have guessed since the words sound similar. Other pnictogens include arsenic, antimony, bismuth, possibly moscovium, and, of course, phosphorus. So basically, it's a family of gases known for their suffocating potential, of which phosphine is a fairly common industrial product used in microelectronics, as a fumigant, and in the preparation of chemicals used in textile manufacturing.

And it all starts with phosphorus. There are two types of phosphorus, at least as far as this discussion is concerned: red and white. These are known as allotropes, or different physical forms of the same element. Diamonds, charcoal, and graphite are all allotropes of carbon, for example.

In the case of phosphorus (P_4), the white allotrope exists as a soft, waxy solid while the red allotrope is an amorphous (or shapeless) solid, meaning it lacks the structure or order seen in crystals; other allotropes include violet and black versions. White phosphorus, which can also appear yellowish, is poisonous and can cause severe burns when coming into contact with the skin. It glows in the dark and can spontaneously burst into flame when exposed to air, which is why it's kept under water. (Note that keeping white phosphorus under water in the lab keeps it stable by preventing a reaction with available atmospheric oxygen.)

White phosphorus is used in flares and incendiary devices, munitions, and military applications like smoke grenades and tracer rounds. It can be manufactured by heating phosphate rock, the naturally occurring form of phosphorus, and then collecting the resulting phosphorus vapors under water.[2] White phosphorus is also the preferred chemical for industrial preparation of phosphine, made by reacting white phosphorus with sodium hydroxide and water,[3] producing a hypophosphite salt [$H_2PO_2^-$] as a byproduct. Using a method known as acid-catalyzed disproportioning of white phosphorus instead yields phosphine and phosphoric acid, the latter of which could have exacerbated the symptoms felt by Emilio and Krazy-8, depending on the strength of the acid.

White phosphorus can be used to make red phosphorus by gently heating it in the absence of air. This nontoxic allotrope is the material in matchbook striker strips, which will come into play during our meth-making discussion. Oddly enough, red phosphorus can be a very effective flame retardant[4] while white phosphorus has been weaponized due to its high reactivity.[5] Red phosphorus can be crystallized upon heating and can produce phosphine gas when reacted with molecular hydrogen. That last fact is exactly what Walt is counting on.

You might be wondering what a jar of phosphorus is doing in a meth lab to begin with. I'll be getting into much greater detail in an entire chapter that deals with the show's various meth-making methods over the seasons, but essentially the phosphorus is used to recycle and regenerate another chemical that's important to the synthesis of meth. In other words, it makes sense that a meth lab would have it on hand, just as it would make sense to find road flares and matchbooks in an illicit meth lab since these are sources of phosphorus.

Speaking of illicit meth labs, let's head to the RV where Walt is cooking up a batch of meth under duress in the episode in question. Though he's going through the motions of his usual cook, Walt's actually stalling until the moment he can throw a jar full of red phosphorus into a pan of hot water and escape during the resulting flash—created by the produced phosphine gas spontaneously igniting in air[6]—leaving Emilio and Krazy-8 to succumb to the deadly poison. Savage science at work.

Except it probably *wouldn't* have worked, not the way this scene portrayed it anyway. The jar of reddish powder at the cook station is clearly labeled "Red Phosphorus," but remember that it's *white* phosphorus that reacts with water vapor and a hydroxide to produce the noxious fumes.[7] Walt only appears to have a pan of hot water on hand in the RV, but even if he'd had some of the hydroxide mixed in, it wouldn't have mattered. Remember that crystalline red phosphorus can react with *hydrogen* to give off phosphine gas. Boiling water does not produce hydrogen gas, otherwise every pot of pasta ever boiled on a gas range would have gone up in flames. (You can "split" water into oxygen and hydrogen gas by running an electric current through water in a process called electrolysis, but clearly Walt's not set up to do that.)

Introducing the red phosphorus to hot water, even with the scalding power of steam, probably would not result in phosphine gas, at least not that quickly. Red phosphorus can react slowly with available oxygen and water vapor to produce phosphine gas, but this reaction is so slow under normal conditions as to not be considered an inhibiting factor in deploying military munitions.[8]

Another way to produce phosphine gas is by introducing calcium phosphide—seen as red-brown crystals to gray, granular lumps—to water. So short of Jesse Pinkman either mislabeling the jar and/or sprinkling some calcium phosphide into it, Walt's ingenious plan to save himself probably would have just gone up in smoke.

Fun Fact: All of the pnictogen hydrides are also odorless in their pure form, but when they come into contact with air, their odors can become quite potent. Nitrogen hydride, better known as ammonia, has a signature smell of fish or urine from the breakdown of urea. Antimony hydride, also known as stibine, smells strongly of sulfur or rotten eggs. Arsenic hydride or arsine has a slight garlic smell, while phosphine smells like fish or garlic. Walt and Jesse would have been well within their rights to complain about a funky smell in the RV after fumigating it with phosphine gas, but Emilio and Krazy-8 had more pressing concerns on their minds.

Side RxN #3: Carbon, by Any Other Name

Let's head back to the classroom to peek in on another of Walt's chemistry lessons. In Season 2, Episode 6: "Peekaboo," Walt is all about carbon. He appears to be speaking another language—"Monoalkenes, diolefins, trienes, polyenes ..."—but at least he admits that getting to know the nomenclature is a bit confusing. He keeps things simple by reminding his class that "Carbon is at the center of it all," and that "There is no life without carbon." For the curious, here's the breakdown of his carbon-based definitions:

- Monoalkenes—Hydrocarbons containing one carbon-carbon double bond (C=C), also known as monoenes. (*Fun Fact*: Walt, or someone in the production, managed to spell this incorrectly as "mono-alkelenes" on the blackboard; the cyclohexene diagram is also missing the double-bond as drawn.)
- Diolefins—Hydrocarbons containing two carbon-carbon double bonds, also known as dienes. An example is 1,3-butadiene, which is used in producing synthetic rubber.
- Trienes—Hydrocarbons containing three carbon-carbon double bonds. (See the pattern?)
- Polyenes—Hydrocarbons containing multiple carbon-carbon double bonds, including dienes and trienes. These can appear in the visible region of the spectrum, like the naturally occurring, yellow-orange polyene, beta-carotene.

Walt goes on to talk about the fact that human beings and diamonds are made of the same stuff, carbon, whether it's a woman's diamond ring on her finger or the woman herself. According to Walt, H. Tracy Hall, "the man who invented the diamond," developed a reproducible process for making synthetic diamonds in the 1950s while working for General Electric. He was only given a $10 U.S. savings bond for his contribution, a process that went on to be used in multibillion-dollar industries.

The late Dr. Hall was indeed part of General Electric's scientific team that made the first synthetic diamond by heating carbon to 5000°F and subjecting it to extreme pressures under a hydraulic press in 1954. This success, which has been repeated over the decades, came from three years of experimentation and hundreds of years of theoretical discussion. Hall's improvements, dubbed Project Superpressure, built on the work of scientists who had come before him, as is often the case. Hall and his colleagues were granted a patent for the process in 1960; Hall followed that up with another patent for an improved press before forming the synthetic diamond-centric businesses MegaDiamond and Novatek.[9] Not bad.

Other reports paint Hall's efforts as more of a solo venture, struggling against GE's refusal to provide him with proper equipment or allow him time to use existing equipment for his experiments. This version of the breakthrough came as Hall worked tirelessly through the Christmas holiday of 1954, only to be accused of exaggerating his claims; his results were then reproduced while Hall was kept well away from the equipment and out of the building. The lack of credit and the $10 savings bond—in addition to his regular salary—proved to be insult to injury, sending Hall to Brigham Young University to continue his research. Hall and his patents were stymied by the federal government on numerous occasions, but he was eventually able to cash in on his research.[10]

Physics

It's not just Walt's vast knowledge of chemistry that gets him into and out of trouble over the seasons but also his general knowledge of other scientific concepts. Take physics, for example, a classic scientific discipline that crosses paths with chemistry quite a bit. A common view of physics is that it exists in the extremes, such as the study of interactions on a very tiny scale in quantum mechanics, and on an astronomical scale in cosmology. But physics impacts our everyday lives in a concrete way, from cooking an egg, to driving a car, to the literal concrete beneath our feet. Physics can also explain why a steel ax blade would never embed itself in asphalt as easily as Season 3, Episode 7: "One Minute" would have you believe. (This scene makes more sense when you realize that it was shot on the studio lot where a square of asphalt was cut out and replaced with a material that looked the part but actually allowed the blade to pierce it and stay upright.[1])

As we discussed in previous chapters, chemistry is a study of intrinsic behavior of matter and systems of matter. Physics, by definition, focuses on *extrinsic* behavior of matter interacting with other bits of matter and the systems in which they operate. Basically, the two disciplines cover the internal and external interactions of all matter. One fantastic example of the real-world interface of physics and chemistry occurs in a very common, everyday object: the battery. I'll revisit Walt's famous DIY battery, which ultimately saved his and Jesse's life in the desert, to see if it really packed enough power to jumpstart their RV.

Building on the understanding of batteries, I'll next talk about the destructive power of electromagnets (especially concerning their effect on a computer hard disk drive) in the hands of Heisenberg. Following that, because I think you might need a drink by this point, I'll be taking a look at the lighter side of *Breaking Bad* and the show's synthesis of drugs you might actually take on a daily or weekly basis: caffeine and alcohol, as consumed after brewing coffee and beer, respectively. If only Walt and Jesse—not to mention Walt's home-brewing, DEA agent brother-in-law Hank Schrader, and Walt's former coffee-addicted lab assistant Gale Boetticher—had opened up a coffee shop or a brewpub, *Breaking Bad* could have joined *Friends* and *Cheers* as the greatest legal drug-based sitcoms of all time.

V DIY Battery

From Dr. Donna J. Nelson:

I received the second question from *Breaking Bad* late one evening: "Using the P2P method, how much meth could be synthesized from 30 gallons of methylamine? And we need the answer in pounds." I found this request humorous for multiple reasons, and it took a while for me to stop laughing. First, during my entire career, I frequently taught organic chemistry to rather large classes of undergraduates, sometimes close to four hundred students. In order to serve as a good role model to those students, I had avoided discussing the synthesis of illegal drugs. Now, to be directly asked to provide information on this topic for television was ironic. Second, scientists typically use the metric system for measurements, and we constantly instruct students not to use pounds. Third, in lab we avoid using large quantities of chemicals for a variety of reasons—safety of small scale reactions, cost of chemicals, cost of glassware to run reactions, impact of the chemicals disposed upon the environment, and so on. I had never used thirty gallons of anything in lab. A request that violated so many of our standard procedures simultaneously was hilarious.

I was unfamiliar with the P2P method and looked it up in the literature (this phenylacetone, or phenyl-2-propanone, method is what Walt uses; P2P shares a basic shape with methamphetamine and Sudafed). There were many reagents that had been used in the reduction step. For the sake of authenticity, I confined my search only to those that had been reported for illicit meth production. There were several, and I knew each would give a different percent yield.

I asked the *Breaking Bad* people if they wanted an exact calculation or only a rough approximation. They wanted it exact. I told them they must identify which reducing agent is used in the final step, so I could look up its percent yield and predict the amount of product obtained. They said they didn't know what all methods were used, so I sent them the list:

- Pd/H_2 catalytic hydrogenation over palladium
- Pt/H_2 catalytic hydrogenation over platinum
- CuO/H_2 catalytic hydrogenation over cupric oxide

- $NaBH_4$ sodium borohydride
- Na/alcohol metallic sodium and alcohol
- $NaBH_3CN$ sodium cyanoborohydride
- Al/Hg aluminum mercury

They said to go with the aluminum mercury because it's relatively simple to pronounce, which would at least keep their actors happy. I also found this humorous; I have selected reagents based on cost, safety, percent yield of product, product purity, ease of procurement, ease of disposal, but never based on name pronunciation. But I wanted this to be a good experience for everyone, so I agreed and proceeded. Nevertheless, I often wondered if the audience suspected how that reducing reagent was selected.

101

Do you have any money? Change, I mean. Coins. ... Gather them, and the washers and nuts and bolts and screws and whatever little pieces of metal we can think of that is galvanized. It has to be galvanized, or solid zinc. And bring me brake pads. The front wheels should have discs. Take them off and bring them to me.—Walter White, Season 2, Episode 9: "4 Days Out"

Batteries. They power our remote controls, cellphones, and laptops. They provide electricity for the starter motors in our cars and the induction motors in modern electric vehicles. They're ubiquitous. And yet, like these devices and their inner workings, of which most people would be hard pressed to sufficiently explain, batteries remain one of the most necessary workhorses of modern civilization that go unnoticed and underappreciated, right up until the moment they fail. And even then, it's much easier (and safer) to simply buy a new battery than to understand how it works and go about making your own.

The closest most of us will ever get to making a DIY battery is the classic schoolroom experiment involving some copper wire, galvanized nails, alligator clips, a low-voltage device like an LED light or clock, and garden-variety lemons or potatoes. This is the earliest example of the inner workings of an electrochemical battery that many of us are exposed to, and it still serves as a great demonstration of the powerful, natural phenomena at work. The zinc in the nail dissolves in the acid inside the lemons or potatoes, producing electrons that travel through the wire connected to the device and then to the copper wire. In this circuit, it's the passage of these electrons through the wire that generates the electricity needed to light the LED or power the clock. Heisenberg simply takes this "potato battery" concept to the next level.

In Season 2, Episode 9: "4 Days Out," Jesse and Walt find themselves stranded in the New Mexico desert when the battery of their RV/meth lab runs out of juice. While this scene serves as a cautionary tale against accidentally leaving your keys in the ignition, there are many, many reasons for a dead car battery in our modern world. The writers of *Breaking Bad* handle this particular situation as if it's not an immediate do-or-die moment for a wannabe drug lord who must resort to MacGyver levels of mad science to save the day, but rather as a common occurrence that usually has a pretty pedestrian solution. You could call roadside assistance for a jump start—assuming you're not cooking illegal meth in a modified RV, which might raise some questions—or, failing that, call a friend to come pick you up. These days, you can even press a button on your smartphone to summon a rideshare from a stranger.

If none of these options work, you'll have to resort to some more mechanical means, like using a gas-powered generator to charge up the battery, or trickle-charging it by hand, or, in very extreme situations, rigging a DIY battery from available materials to give the battery enough of a boost to get the engine going. It's a testament to the strong writing of *Breaking Bad* that Walt and Jesse exhausted every possible solution to the problem of the dead battery before resorting to DIY science to save the day.

After their phones die while potential rescuer Skinny Pete becomes forever lost in the New Mexico wilderness, and after the gas generator catches fire and they use the last of their water supply to put it out, Walt and Jesse are forced to scavenge for parts in order to cobble together ... a robot? No, not a robot, much to Jesse's disappointment. (This hilarious response from Aaron Paul was actually unscripted.[2])

Of course, they're building a battery; specifically, a mercury battery. The one-time-use, nonrechargeable battery is assembled by using on-hand chemicals like potassium hydroxide, as well as co-opted materials like graphite from the RV's brake pads; zinc from galvanized nuts, bolts, screws, and coins; copper wire scavenged from power cords; and mercuric oxide that comes from ... somewhere. (More on that in a moment.)

Walt does manage to assemble all the necessary components of a functioning battery in order to get the electricity flowing, as seen by the spark that jumps from his contraption to the clamp when it's attached, and both the real-world physics and chemistry check out on a conceptual level. But while it doesn't take all that much electrical energy to get an engine going, is Walt's DIY battery powerful enough to actually achieve ignition? To find out, I'll have to go into a bit more detail.

Advanced

An electrochemical battery (known in this case as a galvanic cell) generates electrical energy through chemical reactions that result in spontaneous electron transfer. (This pathway can also go in the opposite direction in an electrolytic cell, in which electrical energy is used to facilitate chemical reactions. The process of electroplating uses this phenomenon to deposit a thin metal coating on a substrate. This will make more sense in a minute.) The galvanic cell consists of two half-cells, each of which is made up of an electrode and electrolyte; the electrolyte may or may not be the same material for both half-cells.

In the case of the fruit and veggie batteries, the electrodes are the copper cathode and the zinc anode, while the electrolyte is either citric acid (lemon) or phosphoric acid (potato). It's important to note that the chemical reactions taking place between the electrodes and electrolyte are the source of energy for these batteries, not the lemon or potato itself. The fruit/veggie flesh does, however, act as a natural salt bridge, a relatively inert medium that maintains electrical neutrality and keeps the reaction running.

Fun Fact: Apparently using boiled potatoes in a potato battery is ten times more powerful than using raw potatoes since boiling reduces the resistance of the salt bridge, allowing the reactions to run more freely.[3]

So why use copper and zinc as electrodes? The answer has to do with the metals' electronegativity, meaning the tendency of an atom to pull electrons toward itself. According to the relative Pauling scale, which measures electronegativity (running from cesium's 0.79 to fluorine's 3.98 on the periodic table of elements), copper has a value of 1.90, making it more electronegative than zinc, which comes in at 1.65. As a general rule, the more electronegative metal will take or accept electrons from the other metal through the conductive wire that connects them, so the copper electrode accepts electrons from the zinc electrode.

Returning to the earlier analogy of using electrons as currency, zinc is a wealthy woman looking to donate some of her hard-earned money to charity; copper is the charity worker who accepts the funds and eventually redistributes them to the needy. The trust that facilitates the transfer of that currency, in this case, is the conducting wire strung between them. (Like most charities, some of that money will go to internal operating expenses so that 100 percent of the contribution won't make it to the other end; conducting wires will lose some energy along the way as heat depending on the wire's physical traits. I'll assume Walt's DIY battery is 100 percent efficient to give him the best possible chance of getting out of the desert.)

So how do the electrons get generated to begin with? Since it is energetically favorable for zinc to give up electrons, this is facilitated by the metal itself actually dissolving in the electrolyte. Positively charged zinc atoms called cations (Zn^{2+}) dissolve into the surrounding solution, leaving electrons ($2e^-$) behind in the anode. Since the anode is connected to the cathode via a conducting wire, the free electrons are able to travel through the more electronegative copper wire to the copper electrode. Here, the electrons are available to interact with available hydrogen atoms (H^+) in the acidic electrolyte on the surface of the copper cathode. Hydrogen (H_2) gas is formed, which bubbles away, also explaining why even traditional batteries generate hydrogen gas during charging and can pose an explosive fire hazard.

In batteries that use a copper-based electrolyte, some copper cations (Cu^{2+}) in solution are also able to interact with the available electrons, causing them to precipitate onto the cathode as copper deposits. In other words, the zinc electrode dissolves while aqueous copper in solution is deposited onto the copper cathode. You can actually weigh the electrodes before and after allowing the battery to run and you'll see a reduction in the zinc anode's mass while the copper cathode gains mass. And remember that salt bridge and electrolyte solution I mentioned? They allow for a flow of the dissolved ions to maintain equilibrium, which keeps the reactions chugging along.

That's the chemistry side of things, so where does physics come in? Remember that free electrons are generated from the anode, conducted along the wire, and used to neutralize cations at the cathode. If you place something between the two ends of that conducting wire—something like an LED light, a clock, or even an engine's starter motor—you can generate an electric current and put those electrons to work, all thanks to the intermingling of chemistry and physics in action!

Just how much work you can get out of a given battery, however, is dependent on something called "standard electrode potential." In common speak, this is another term for voltage. A dry cell battery, like a standard AA, uses a zinc anode and a carbon cathode and has a voltage of 1.5V (volts) ... but so does the smaller AAA battery. This is because voltage is determined by the chemicals and materials that make up the battery, not the size of the battery itself. Primary, non-rechargeable zinc-carbon batteries have a voltage of 1.5V while rechargeable nickel-cadmium/metal hydride batteries produce 1.2V. A mercury battery, meanwhile, has a steady voltage discharge of 1.35V, making Walt's DIY battery a strong candidate to replace the twelve volts that a functioning RV battery would have provided.

So now that we have a solid understanding of the basic components of a battery and how its chemical reactions generate electricity, let's take a look at Walt's attempt to jump-start his RV.

For the anode, Walt asks Jesse to gather up coins, washers, nuts, bolts, and screws, as long as they are galvanized or solid zinc. The oxidation potential of this reaction—one-half of the overall standard electrode potential—is +0.763V.[4] (Galvanization is the process of applying a protective zinc coating to steel or iron to prevent oxidation, or rusting. Walt was probably better off sticking with the hardware instead of coins since U.S. currency is actually a copper and nickel—or cupro-nickel—composition; post-1982 pennies are a good source of zinc once their copper coating is scratched away, however. A stack of alternating pre- and post-1982 pennies can actually be used to make a small copper/zinc battery!)

For the cathode, Walt uses available mercuric oxide—a.k.a. mercury (II) oxide—on hand. I'm still unclear about whether he had this chemical available from his meth-making business—it would make some sense considering their cooking method, which I'll discuss later on—or if it's assumed to be a component of the ground-down brake pads; the show itself is unclear about the origin. While modern brake pads do have some metal oxides to increase friction, I've yet to find a brake pad in which mercuric oxide is used since various iron oxides seem to be preferred, along with zinc, aluminum, and magnesium oxides. Regardless, the reduction potential of the cathode is +0.855V, if we're assuming that the mercury is available as Hg^{2+} in order to give Walt's battery the best chance of working.[5]

What is clear, however, is that Walt obtains the graphite (crystalline carbon) from the ground-down brake pads; the material acts as a lubricant. In Walt's battery, the graphite helps to facilitate electrons to the mercuric oxide since HgO on its own is a nonconductor; it also prevents the pooling of toxic, elemental mercury (Hg) that is produced during the reaction.

The final components are as follows: aqueous potassium hydroxide (KOH in water) as the battery's electrolyte, a sponge soaked in said electrolyte to act as the salt bridge, copper wire used to conduct electrons, and jumper cables used to connect the DIY battery to the RV battery. All of these materials would be readily available in the RV meth lab.

In Walt's battery, the zinc anode is oxidized to zinc oxide (ZnO) in the KOH solution and loses two electrons, which travel through the copper wire—and connected RV battery, theoretically charging it—to the mercuric oxide/graphite cathode, where HgO is reduced (gaining those two free electrons) to form elemental mercury (Hg). The overall voltage potential of each cell is, at maximum, 0.763V + 0.855V = 1.618V. Walt and Jesse had enough material for six cells, meaning that the total available voltage is 6 × 1.618V = 9.708V, without taking inefficiencies like voltage loss, electrical resistance, and minor side reactions into account. It's a worthy little electrochemical powerhouse considering that it was cobbled together from items on hand, but is it enough to jumpstart the RV battery?

Sadly, probably not. There are a few things stacked against Walt's DIY battery here. The first is the available voltage. Even though I used a very favorable reaction potential for the HgO, the chemistry actually suggests the value would be 0.0977V, slightly more than a tenth of the value I'm generously giving to Walt's battery. But even at the maximum theoretical voltage, it's still less than the RV's assumed 12V battery.

Attempting to jump-start a dead 12V battery with a battery of a different voltage—DIY or otherwise—is dangerous and potentially explosive, though usually for the battery with the lower voltage. (Remember that charging lead-acid batteries can also produce hydrogen gas, adding an explosive fire hazard to the DIY setup.) But perhaps the 1986 Fleetwood Bounder RV didn't have the standard 12V battery ... perhaps it had two 6V batteries used together in series, and perhaps only one of those was dead.

That's a lot of "perhaps," but even if that were the case, Walt's DIY battery probably didn't have enough current to charge the dead battery. If we think of electricity in terms of water (a rare scenario in which these two should mix), voltage can be understood in terms of pressure. Current is the "volume per unit time" or flow of electricity from high voltage to low voltage (note the difference in the two batteries' voltages here) and is equal to the proportion of voltage to resistance, as Ohm's law demonstrates.

Using the water metaphor, resistance is basically the size of the pipe through which the water flows; a pipe with a larger diameter has less resistance and vice versa. The copper wire Walt used, while a good conductor, was probably too small in diameter to allow enough current to flow between cells and into the jumper cables on its way to the dead battery. And while potassium hydroxide cells are good for providing constant voltage at higher currents than, say, sodium hydroxide, the maximum amount of current cranking out of the DIY battery was probably insufficient, far short of the minimum 500 cold-cranking amps (CCA) needed to jump a battery.[6]

So we've got a lower-voltage battery attempting to charge a higher-voltage battery (unless we swap a dead 6V battery in place of the likely 12V battery), the conducting wire from the battery itself is relatively thin compared to jumper cables, and the available current is less than a tenth of the minimum necessary to jump a dead battery. It's not looking great for Walt and Jesse and their DIY desert rescue. Maybe next time they should look for an RV with a manual transmission and a carburetor, which can be push-started in the case of a dead battery ... or just charge the phone instead ... or maybe make sure Jesse doesn't leave the keys in the ignition.

Fun Fact: Historically, mercury was included in button-cell batteries, either in the anode itself or the paper insulation surrounding the battery as a way to prevent zinc corrosion, which leads to a build-up of hydrogen gas and a decrease in the battery's function. These batteries were banned in the United States in 1996 due to mercury's toxicity, though some large mercuric oxide batteries are still used in military, medical, and industrial applications.[7]

Inside *Breaking Bad*: In Season 1, Episode 4: "Cancer Man," Walt ruins the BMW driven by Ken, an obnoxious stock broker played by Kyle

Bornheimer, by placing a squeegee between the battery terminals of his car. The fiery result was actually inspired by Vince Gilligan's own brother who accidentally shorted out his car battery in their driveway when they were younger. As Gilligan says, a car battery can put out hydrogen gas upon charging—in addition to the dangers posed by sulfuric acid in the battery itself—and that gas could ignite under the right conditions. The Gilligan brothers found this out the hard way when the short-circuited battery exploded. Luckily, Vince's brother escaped without injury by sheer luck, though I can't say the same for Ken's Beamer and its vanity license plate, KENWINS.[8]

Side RxN #4: *Breaking Bad* Balloons

In the Season 3 finale, "Full Measure," fan-favorite fixer Mike Ehrmantraut heads to a warehouse where cartel hitmen have taken a chemical supplier and his secretary hostage. Mike's neutralization of said hitmen is brutal in its simplicity, but it's his clever method of knocking out the building's power that had me doing a little extra digging.

After giving his granddaughter a bunch of helium-filled Mylar balloons, Mike oddly keeps a dozen or so for himself. That night, he approaches the warehouse from a safe distance and lets them fly. They rise until they bump into the overhead power lines, which arc and spark until the transformer blows, knocking out the power—and the security cameras—to the building in question. The rest is just wet-work.

Believe it or not, metallic and Mylar balloons pose a serious hazard when it comes to power lines. Careless drifting of these otherwise harmless party favors can "disrupt electric service to an entire neighborhood, cause significant property damage and potentially result in serious injuries" when coming into contact with the lines. Mylar balloons bumping into power lines creates a short circuit—a short circuit being the flow of electric current along an unintended path—according to the Pacific Gas & Electric Company, among other utility companies.[9] There are plenty of videos of this explosive, destructive phenomenon online, all of which seem to be accidents or instances of carelessness rather than setups for an assassination. Kudos to the *Breaking Bad* team for including this explosive idea.

Inside *Breaking Bad*: For years, Gilligan had a vision for an anecdote featuring a tough guy carrying a bunch of balloons, but he never had a way to work it into a scene or script. He had also randomly come across the factoid about Mylar balloons and their hazards when it comes to electrical lines. Those two ideas came together to make this scene happen.

As a bonus bit of trivia, the logo on the methylamine drums stored here and seen throughout the series comes from Golden Moth, Inc. The show's art department designed signage for the company with both English and Chinese characters. The logo was supposed to be shown on the side of this chemical supply building but due to production issues it didn't make the cut.[10]

VI A Magnetic Conversation

From Dr. Donna J. Nelson:

No script pages accompanied their second request for assistance, so I wondered why they specified thirty gallons. A quick look at commercial availability of methylamine in large quantities revealed that this is a standard-size drum for that compound. Still, I didn't immediately understand why they would want to use such a large amount. We avoid using large amounts of chemicals in university labs. However, it becomes obvious in the scene where Walt and Jesse break into a locked storage room in order to steal the methylamine needed for their next synthesis. They originally intend to take only one gallon of $MeNH_2$, but there are no one gallon containers; they find only thirty-gallon drums so they take one of those.

This scene is important for another reason: they use thermite to break into the locked storage room. But why would the writers use that method of entry? Usually a burglar would simply knock off the lock with a hammer. However, they wanted science in the show to be bigger than life or spectacular; it often outshines the actors, as it does in this scene. Imagine this scene as they prepare to break in—Walt and Jesse are dressed in black. It is nighttime, and everything behind them is dark. Then they ignite the thermite and sparks sail across the screen; the sparks obscure everything else, and they are all the viewer can see.

In *Breaking Bad,* the chemistry frequently takes over the scene. Think about the scenes in which the bathtub falls through the ceiling because the wrong container was used, the entire building is blown up with a large crystal of mercury fulminate, the elaborate coffee brewing equipment is used and discussed at length, and there are many others. This is what sets *Breaking Bad* apart from every other TV show with science content. The science is often so strong that it becomes a separate character. During a set visit, I asked Vince if he had intended for the science to be so strong that it seemed to be a character in the show; he smiled and simply said, "Yes."

Also on one of my set visits, in talking to Vince about the methylamine, I referred to it as a precursor in the reaction. He became excited and said, "Precursor? What is a precursor?" I explained, and thereafter, I noticed that the word "precursor" was used frequently in the show. I suggested that if he really wanted to impress the audience with scientific words, he should incorporate "stoichiometry" or "stoichiometric" when discussing the amount of product synthesized. After all, the calculation of how much meth would be generated from thirty gallons of methylamine is a stoichiometric calculation. However, he wasn't interested. I assumed one reason was that the pronunciation would have been difficult for the actors, a factor that had also influenced their selection of aluminum mercury for the reducing reagent.

101

Yeah, bitch! Magnets!—Jesse Pinkman, Season 5, Episode 1: "Live Free or Die"

In the Season 5 premiere, "Live Free or Die," Walt and Jesse have finally managed to get out from under the thumb of meth mogul Gustavo "Gus" Fring after a devastating pipe bomb explosion (arranged by Walt) takes his life. Fring, the owner of Los Pollos Hermanos chicken restaurant chain, which masked his drug distribution system, was unable to save face when his criminal activities were posthumously revealed, though Walt and Jesse had yet to be implicated. However, some security camera evidence from Gus's Superlab/laundry facility still exists on a laptop hard drive, a laptop that just so happens to be in the possession of the Albuquerque Police Department. While Walt argues with Mike over just how secure the APD's evidence locker really is, it's Jesse's suggestion of using a magnet to destroy said evidence that ultimately gets the gang cooperating once again. But how it all plays out isn't exactly what Jesse, Walt, and Mike intend.

A classic science demonstration of magnetism uses iron filings (tiny pieces of iron that appear as a powder) in the presence of a simple bar magnet in order to visualize its magnetic field. It's that invisible field that allows for interactions between magnetic objects. Traditional computer hard drives use this phenomenon of magnetism to store information by magnetizing or demagnetizing billions of tiny little sectors, represented in binary as 1s and 0s. So it's conceivable that a powerful enough electromagnet could wipe out this delicately stored data.

Utilizing a massive electromagnet courtesy of Old Joe, who uses it to move cars in his junkyard, Walt and Jesse rig up enough batteries (not the DIY kind, thankfully) to crank the magnet's power to the max outside of the evidence locker. This not only, presumably, wipes the hard drive's magnetically stored information, it also destroys every magnetic object in the place by smashing them against the evidence locker wall. But while the science of the battery-powered magnet itself is surprisingly sound in this episode, Walt and Jesse's plan would have likely run into some electromagnetic interference in the real world. It turns out that modern hard-disk drives are reassuringly difficult to demagnetize, even with a supervillain-level electromagnet on hand. However, the tried-and-true, brute-force method of destruction could have rendered the laptop unusable anyway.

To better understand why Walt and Jesse's mobile magnet might have worked but ultimately would have come up short in terms of erasing any data on the hard disk drives, we'll have to get into the specifics of electromagnetism and computer data recording and security.

Advanced

Magnets: How do they work? A magnet is defined as any material or object that produces a magnetic field. The most notable effects of this invisible field are the attraction of ferromagnetic materials to the magnet and the ability to attract and repel other magnets. The Earth itself is a giant magnet thanks to the rotating, convecting, and electrically conducting material of the planet's core, according to the dynamo theory. This massive magnetic field not only protects the Earth's atmosphere from solar winds that would otherwise strip away the ozone layer, it also provides us with an incredibly useful navigational tool. As a bonus, it gives us a method to track the ancient motion of continents and reversals in the polarity of the field itself thanks to magnetic evidence recorded in igneous rocks. That's a lot to wrap our heads around, so luckily, we can also explore the effects of the magnetic field on items as small as refrigerator magnets or, of course, on the scale of car-lifting electromagnets.

But before we get into the specifics of how an electromagnet works, it's worth revisiting how ferromagnetic materials like iron, nickel, cobalt, and rare-earth metals can be magnetized to generate a magnetic field. Known as permanent magnets, these materials can be magnetized in a number of ways:

- Heating them above their Curie temperature—the temperature at which the material loses its magnetic properties—and then hammering them as they cool within a magnetic field. This is the most effective method, a simplified version of which is used in industrial processes.
- Placing them in a magnetic field and applying vibration, which allows them to retain some residual magnetism.
- Stroking an existing magnet on a piece of ferromagnetic material and/or applying an electric current to it can also create a permanent magnet.

These materials can be further divided into magnetically "hard" and "soft" categories, depending on their natural tendency to stay magnetized. Magnetically "hard" materials, like the iron alloy known as alnico and the ceramic compound ferrite, tend to stay magnetized, while "soft" materials like annealed (heat-treated) iron do not.

Things can get very complicated when talking about magnetism on a subatomic level, but the simplest version is this: Ferromagnetic materials have a unique arrangement of electrons that makes them more likely to align with a magnetic field. This explains why certain materials, like Walt's gold rings and nonferrous glasses, are not magnetic, or at least not affected by magnetic fields on any appreciable level outside of using highly sensitive laboratory equipment.

In more detail, electrons in an atom's orbital have properties that are represented by quantum numbers. These define an electron's energy and spin (one of two directions, either up or down), and the orbital's size, shape, and orientation in space. Movement of electrons in these orbitals creates a very small magnetic field. These electrons tend to orbit in pairs, and due to the Pauli exclusion principle—which, as noted earlier, states that no two electrons can have the same quantum numbers—their spin values will be opposite each other and their magnetic fields will cancel each other out.

Ferromagnetic materials, however, have partially filled electron shells occupied by unpaired electrons that have the same spin; their magnetic fields are not canceled out. This allows for an orbital magnetic moment, a vector that has both magnitude and direction and influences nearby atoms to align in the same north-south orientation within the magnetic field. When these ferromagnetic materials cool, after being heated either by natural processes or industrial ones, the atoms with the same magnetic moments line up within the crystalline structure, giving rise to magnetic domains.

If all that talk of particle physics has left your head spinning, don't fret. The main takeaway is that electric current and magnetic fields are all wrapped up together. That's a good thing, otherwise Walt's plan to use the electromagnet would have short-circuited before it began.

In an electromagnet, the magnetic field is produced by the flow of electric current; it's all right there in the word itself. That means that a magnetic field can be found around anything from a single straight wire, to a solenoid, to a toroidal—or ring-shaped—coil, as long as current is flowing. Commonly, that current flows through a coil of insulated wire (a solenoid), which can also be wrapped around a ferromagnetic material to increase the strength of the magnetic field. If the current is amped up, the strength of the magnetic field also increases, according to Ampère's law. Using this basic concept, engineers can make electromagnets for use either in magnetic storage devices in consumer computer parts or industrial magnets used to lift scrap iron in a junkyard, both of which feature heavily in this episode of *Breaking Bad*.

But before we lose our current train of thought, let's take a moment to talk about Walt's insistence on a second string of car batteries used to power the electromagnet. You should have a basic idea of how a battery works from chapter 5, so I'll kick that understanding up a notch for this discussion. Initially, Old Joe has wired up twenty-one 12V car batteries to power his repurposed junkyard electromagnet. Because these batteries are wired in series—meaning that a wire connects each battery's negative terminal to the positive terminal of the next battery in line, similar to how AA batteries can be lined up end to end in a small flashlight or wireless keyboard—their voltages all add up to a total of 252 volts. That's more than enough to supply the 230 volts required by the industrial electromagnet, as seen in the gang's test run.[1]

However, Walt also wants another twenty-one of those same batteries wired in parallel. Rather than being connected end-to-end in series (negative to positive), which increases the available voltage, batteries wired in parallel (positive to positive, negative to negative) doubles the *capacity*, measured in amp-hours. This ups the available current for the electromagnet. And as I already mentioned, the strength of a magnet's magnetic field increases or decreases proportionately to the available current. Basically,

Walt wants to double the available strength of the electromagnet to make sure it gets the job done.

But what is his endgame here, exactly? Is Walt just aiming to cause enough chaos in the evidence room by smashing all available ferromagnetic materials inside it in the hope of destroying the laptop? Or is he really trying to wipe the computer's magnetically stored information? While the brute force method is a nice back-up plan, the magnetic erasure is the cleverer approach ... but would it work?

Modern computers still use magnetic recording—the process of storing data on a magnetized medium—to store data on hard disk drives (HDD), even though they were first invented over sixty years ago. Granted, every aspect of the modern HDD is vastly improved compared to its original ancestor. But the moving parts of the devices, which are necessary to read and write data, remain as possible points of mechanical failure under normal use, not to mention the possibility that a mad scientist/drug kingpin will attempt to use powerful electromagnets to wipe an HDD of all its painstakingly recorded data. If Gustavo Fring had used flash memory in a USB stick or solid-state drive (SSD) instead, Walt would have had a tougher time destroying the information on these nonvolatile memory storage devices.

So let's assume Gus's laptop had the more common and more traditional HDD in which the security camera video was stored. Would it be wiped out by a powerful electromagnet as easily as a credit card's magnetic stripe would certainly be? (Old Joe did warn everyone to keep their cards far away from the magnet's area of effect if they "want that plastic workin' come Miller time.")

The way HDDs read and record data is through the use of a read/write head at the end of an actuator arm. As this head moves over the magnetic surface of the disk—with a separation of as little as 3 nm (nanometers)—it comes across very small (submicrometer) regions known as magnetic domains. During the writing of data, the head magnetizes a region by using available electric current to generate a strong local magnetic field; whether a region is magnetized or not represents 1s and 0s, the bits of a computer's binary code. During reading, the same head detects whether a given region is magnetized or not, with the help of a controller. With all of these tiny parts, high speeds, and sensitive magnetic manipulations, you might think that a strong electromagnet would be more than enough to completely scramble a laptop's HDD.

Well, the folks at K&J Magnetics, Inc. put one of their own HDDs to the test. After pulling a 30GB (gigabyte) HDD out of its computer housing and loading it up with a repeating line of text, the techs placed very powerful neodymium magnets on either side of the spinning disk.[2] Care was taken not to place the magnets over the read/write head since they were aiming to disrupt the data itself, not the tool used to read and store it. In other words, this setup was even more likely to cause a data failure than Walt's scheme since the magnets in question were much closer to the HDD. However, 100 percent of the files were intact and accurate after their test.

Part of the reason for this is that the HDD material's coercivity, or resistance to being demagnetized, was sufficiently high enough to prevent erasure. This erasure, called degaussing, can only be achieved if the degausser's magnetic field strength is two to three times the coercivity of the HDD material, measured in oersteds.[3] (Anyone who has used older, cathode ray tube [CRT] TVs or computer monitors may remember using a degaussing function to "wobble" the screen. This function oscillated the tube's magnetic field, somewhat randomizing it, and removed any discoloration due to picking up strong external magnetic fields.)

Another factor at play here is the HDD's housing, which shields it somewhat from the magnets; the outer laptop housing would provide additional protection. Clearly, Walt couldn't get his hands on the HDD to send to a professional degausser, nor could he get access to the laptop itself to run a data-wiping program, and the electromagnetic wipe of the laptop's data probably wouldn't have worked despite the impressive setup. The next best option: Brute-force destruction.

While the incredibly strong rare-earth magnets failed to corrupt the HDD data in the real-world experiment, they easily could have warped the drive's spinning platter and sensitive components like the actuator arm and read/write head. Even without the added stress of powerful magnets, HDDs have been known to fail during the process of reading/recording data, which is normally caused when the read/write head comes into contact with the platter—a catastrophic failure known as a "head crash"—or due to an incorrectly moving actuator, producing the "click of death." And if those relatively small-scale destructions of the computer's components weren't enough, sending every ferromagnetic object in the evidence locker smashing against the steel-reinforced concrete wall could have achieved a sufficient level of catastrophe to destroy the computer completely.

But did it all actually work? Well, there was a follow-up shot of the evidence locker showing Fring's busted-up Samsung laptop computer, but as for whether or not the HDD was still functional, you'll just have to take it on faith and trust in Heisenberg that the evidence was erased. Why? Because he said so.

Inside *Breaking Bad*: The Hollywood side of things is pretty interesting for this one. For starters, Mike had to blind the facility's security cameras before any magnet madness could take place. To do this, he used a can of wasp or hornet spray due to its long range (twenty feet). Gilligan confirmed as much on the related podcast episode.[4]

Now, back to the magnets. The idea of using an industrial-sized degausser came, in part, from Gilligan himself who used a similar device to erase audio tracks during his film studies course at New York's Tisch School of the Arts.[5] But as for the physics of magnetic fields, that can get complicated rather quickly, as seen in equations like the Biot-Savart Law, which is used to calculate magnetic field strength:[6]

$$B = \int \frac{\mu_0 I}{4\pi r^2} \, dl \times \hat{r}$$

where,

- B is the magnetic field density (a vector)
- dl is the differential element of the wire in the direction of conventional current (a vector)
- r is the distance from the wire to the point at which the magnetic field is being calculated
- $\hat{r}$ is the unit vector from the wire element to the point at which the magnetic field is being calculated

However, the episode in question did a fantastic job of making the scientific concepts accessible without getting bogged down in the details. Take the rate of decay of a magnetic field's strength over distance, which arguably decreases either at an inverse square rate or an inverse cube rate, depending on a number of factors. The show skirted this discussion with a simple visual demonstration, showing that a laptop would fritz out completely at a distance of about twenty-five feet from the powered-up electromagnet before zipping through the air to smash against it.

Of course, this effect was done practically through the use of cables pulling the laptop out of Jesse's hands to send it flying into the side of the truck; the cables were digitally removed later on. And while the computer monitors in reality may not have become pixelated as an early warning sign that some strange attraction was at work, at least the episode established the electromagnet's effective range before putting it into action at the APD.

But this wasn't the only scene of magnetic trickery in "Live Free or Die." Even though the electromagnet was a legitimate one used to move scrap iron, a behind-the-scenes featurette from AMC revealed that special effects technicians had to add two inches of magnetic plate metal weighing 1,500 pounds to the underside of the junkyard car's roof to provide enough mass for the magnet to lift it.[7] Normally, these electromagnets move cars that have already been crushed down into a nice, compact approximation of a cube, making them much easier to lift and move.

The electromagnet seen in the box truck, however, was a fabrication, a fifty-pound copy made of foam to stand in for the three-and-a-half-ton iron workhorse needed at the junkyard. And while some of the exterior shots at the APD's evidence room were filmed at a real substation, the wall that the truck eventually leans against, the interior of the room itself, and the procedures that the brave men and women of the fictional APD follow to handle evidence were all just smoke and mirrors (and a little glue and plywood) used for the show.

Side RxN #5: Zombie Computers

When it comes to computer science, AMC's show *Halt and Catch Fire* dances circles around *Breaking Bad*. However, there's another subtle addition in the show that reveals just how clever and connected Walt's lawyer Saul Goodman actually is.

In "Phoenix," the penultimate episode of Season 2, Walt Jr., Walt's high-school-age son, shows off his charity website SaveWalterWhite.com. (This narrative idea actually grew out of an earlier idea that Skyler, Walter's wife, was using eBay to sell things and make a little extra money.[8]) Intended as a way to crowd-fund Walt's surgery as part of his cancer treatment, the website also gives Saul the perfect avenue for laundering Walt's ill-gotten gains. Saul's "hacker cracker" in Belarus sends small, anonymous, and consistent donations to the site by turning domestic desktops into "zombie" computers. It works, despite Walt's prideful issues with the plan, but is it realistic?

Well, back in 2008, Homeland Security Secretary Michael Chertoff thought the threat was real enough to address it during that year's RSA Conference, an IT conference organized by RSA Security.[9] While money laundering is certainly a concern, the more pressing cybercrimes include spam, credit card theft, and distributed denial of service, or DDoS, attacks, intended to take down rival sites, extort their owners for money or information, make political and ideological statements, or even attack critical infrastructure. Botnets—networks of privately owned computers controlled by other individuals through malicious software without the owners' knowledge or permission—continue to be a problem today, but it's doubtful that SaveWalterWhite.com would have been convincingly able to launder the tens of millions of dollars Walt eventually made.

Fun Fact: The website SaveWalterWhite.com still exists today and its donation link is active. However, it links to AMC's site for the show; it used to link to a cancer charity until it was ranked as one of the worst charities in America.[10]

VII Trouble Brewing

From Dr. Donna J. Nelson:

Back to the question on stoichiometry; how much meth will thirty gallons of methylamine make?

This question would be excellent to include in homework for a general chemistry course—or as a review question in organic chemistry homework. The subject would probably be too distracting for it to serve as a test question in either course. It is not a simple stoichiometric calculation, because additional information about the reactants and reagents must first be obtained.

First, one needs the percent yield of the reaction using the reduction method that the *Breaking Bad* writers selected—Al/Hg. Although Al/Hg (aluminum mercury) was selected because it was easy for the actors to say, information about it was relatively obscure, which made this the hardest part of the stoichiometric calculation. A thorough literature search revealed that in 1964, Al/Hg reduction was patented in Germany for commercial production of methamphetamine in 70 percent yield (i.e., 70 percent of the theoretical maximum). Yes, the patent is written in German, which could pose an additional level of difficulty for some scientists; fortunately, the school where I obtained my PhD (UT-Austin) required two semesters of scientific German in its PhD program.

Second, the assumption is made that Walt, with excellent lab technique due to his PhD from the California Institute of Technology and access to excellent equipment, matches the maximum yield reported in the literature for both steps in his reaction. This will be a quantitative yield (approximately 100 percent) for the first step and 70 percent for the second step in his synthesis, as is reported in the literature.

Third, a thirty-gallon drum of $MeNH_2$ is a solution of only 40 percent $MeNH_2$ in H_2O, as available from a commercial source, such as is depicted in the show.

The calculation then is as follows:

- 40 percent of 30 gallons of $MeNH_2$ = 45.4 liters $MeNH_2$
- (45.4 L) (0.66 Kg/L) = 30 Kg = ~967 moles $MeNH_2$
- With 70 percent yield, the above gives ~677 moles of meth

- = 101 Kg (but the answer was requested in pounds)
- = 223 lb of *d,l*-methylamphetamine

I mentioned to the writers that about half is wasted because only 50 percent of the racemic mixture is highly active in the human body, but they didn't care about this.

After spending effort on the calculation, I was interested in its impact on the show; considerable attention was paid to determining the quantity of drug synthesized. This attention to detail is typical of the effort put into having accurate science throughout the *Breaking Bad* series. However, the impact of the calculation of amount of product on the script dialogue was minimal. In one clip, Walt uses the value (223 lbs) to calculate that he won't need to "cook" "for the foreseeable future." In a second, brief clip, his brother-in-law, DEA Agent Hank Schrader anticipates the effect of a large amount of meth upon the drug community, stating, "Thirty gallons of precursor—that big a score, they're going to wind up stepping on some toes."

101

See, in my opinion, it's all about the quinic acid level. You want just north of 4,800 mg per liter, but if you over-boil to get there, you're gonna leach your tannins ... bitterness, yuck. So I pull a mild vacuum that can keep the temperature no higher than 92°C.—Gale Boetticher, Season 3, Episode 6: "Sunset"

Not everything in *Breaking Bad* is centered on the synthesis, consumption, and dealing of illegal substances. In fact, some of the drugs present in the show are quite common; you might even have some of them in your own house right now. Caffeine—as delivered by everyone's favorite morning beverage, be it coffee or tea—helps to keep us alert and get us through the workday. Alcohol, on the other hand, helps the mind and body to relax (when taken in moderation, of course) after a stressful eight-hour shift or forty-hour week. These are both drugs by definition. But unlike methamphetamine, we can make, consume, and trade both caffeine and some kinds of alcohol without fear of Johnny Law kicking down our doors.

There are two characters in *Breaking Bad* who have very subtle side plots when it comes to synthesizing perfectly legal drugs. One concerns Gale Boetticher, a very competent organic chemist with a specialization in x-ray crystallography who becomes Walt's lab assistant for a short time; it's his side quest to brew a perfect cup of coffee that added an extra wrinkle to his character.

Inside *Breaking Bad*: The character of Gale Boetticher was named after the late, Oscar-nominated director Budd Boetticher who helmed the 1951 picture *Bullfighter and the Lady* and Westerns like *Ride Lonesome* and *Comanche Station.*[1] And although David Costabile was fantastic in the role of Gale, Gilligan originally had the recently deceased Philip Seymour Hoffman in mind.[2]

DEA Agent Hank Schrader, meanwhile, makes his living by busting drug dealers on every level from the local street corner to international cartels. In his downtime, he likes to blow off steam by home brewing his very own beer, Schraderbräu. But while Gale's coffee habit doesn't land him in hot water, Hank's overpressurized home brew just about gives him a heart attack.

There are actually some solid real-world scientific parallels between Gale's method for the perfect cup of coffee and even Hank's super-carbonated bottle bombs, so we can file these side stories away under the category of "Mostly Right." The brewing processes of coffee and beer are among the most common examples of everyday science we have, so I won't bore you with the step-by-step process for each here except to say that they both rely on using hot liquids to extract desirable compounds from plant-based materials, be they coffee beans or malted barley and hops bound for beer production. That sounds a bit clinical, especially since some consider brewing of either potent potable as more of an art than a science, but an understanding of the many scientific disciplines that interact in the brewing process will take you a long way toward making the perfect cup or pint.

But what about Gale's overly complicated coffee-brewing setup? And what exactly caused Hank's extra-effervescent beverages to explode? Were they practical issues, or a result of Hollywood dramatization? I'll turn once again to physics to find out.

Advanced

When it comes to the perfect cup of coffee, everyone's a critic and everyone's an expert. (Grab yourself a mug of the good stuff while reading this if you want the full sensory experience.) The matter is subjective, so a homemade cup of instant Joe, a pot of your local diner's finest coffee (black, of course), or the most dessert-like beverage served up at a convenient franchise location could be your perfect cup of coffee. But one unpalatable aspect of coffee across the board is bitterness.

Bitterness is a part of coffee's overall flavor profile even in a good cup. It moderates acidity and adds complexity when present at low levels; at higher levels, however, it can overpower the other flavors. (We humans are quite sensitive to bitterness, evolutionarily speaking, since the flavor can indicate a possible poison.) It's perceived by the interaction of taste buds with the bitter chemical in question at the back of the tongue, and it just so happens to be a byproduct of coffee-making itself, a process otherwise known as extraction. "Extraction" is a more scientific-sounding term that describes the mixing of coffee grounds in hot water in order to dissolve their flavor compounds into a drinkable brew, a common process performed daily in everything from a traditional French press to modern, high-tech coffee machines. It's *over*extraction that tends to pull out more of coffee's naturally occurring bitter compounds, such as caffeine, trigonelline, furfuryl alcohol, and quinic acid, to name a few. It's that last chemical that Gale Boetticher was so concerned with.[3]

There are a few ways you can reduce the bitterness in your everyday coffee without resorting to Gale's overly complicated setup:

- To limit overextraction of bitter compounds, once the coffee is brewed it should be kept separate from the grounds, as in a French press.
- Similarly, a fine, even grind allows for more of the coffee beans' compounds to be extracted, but since the bad comes along with the good, too fine a grind increases bitterness.
- Use the Goldilocks method for your brewing water temperature: too hot and you'll extract too many bitter compounds, but too cool and you'll miss out on the extraction of aromatic compounds that counter coffee's bitterness. Aim for 195°F to 205°F for optimal extraction.[4]

Speaking of water, an easily overlooked but essential part of the brewing process, it's better to use hard or soft water than distilled since water's mineral content—particularly magnesium—aids extraction and improves flavor.[5] Your preference may certainly vary by roast, but if bitterness is a problem for you, try a medium roast; decaffeinated coffee is also less bitter but that's understandably a nonstarter for many coffee drinkers. Other quick fixes include making sure that your coffee-making equipment is clean and residue-free; using a drip or siphon system instead of a French press; using fresh, whole beans; and, of course, putting a little sugar in your cup to sweeten things up.

But let's assume Gale, fastidious as he was, already knew all of these tricks of the trade. He seems to, anyway, saying, "Sumatran beans. And I also have to give credit to the grind," after Walt compliments his brew. Gale was concerned with one final stumbling block when it came to making the perfect cup of coffee: quinic acid.

As I previously mentioned, quinic acid is a component of coffee that's partly responsible for perceived bitterness, as well as its acidity and astringency (often confused with bitterness, astringency is actually a dry, puckering mouthfeel). It's a member of the Tara tannins, so named for their existence in the Tara tree (*Caesalpinia spinosa*) that grows wild in South America and North Africa. Gale's recommended concentration of quinic acid falls in line with the 3,200–8,700 mg/L (milligrams per liter) range found in roasted coffee; the taste threshold is just 10 mg/L, however, so it doesn't take much for the chemical's presence to make itself known.

So Gale wanted to finely tune the amount of quinic acid present in his final brew without throwing off the balance of other flavor compounds. His comment about overboiling, in other words, heating the water to a boil (100°C or 212°F) and allowing it to cool slightly for the actual extraction, makes sense since too hot a temperature would overextract unwanted tannins. Ideally, you'd want your water to boil at a lower temperature in order to achieve the best of both worlds, but that goes against the known laws of physics, doesn't it?

Not at all! One of the most useful aspects of a thorough understanding of scientific concepts is that you can tweak one variable or another in order to change the outcome. Gale took advantage of his understanding of physics to make his perfect cup of coffee by dropping the pressure in the closed brewing system to generate a vacuum, allowing the water to boil at a lower temperature. The reason this works is due to vapor pressure, which is the pressure exerted by a material's gaseous phase on its liquid or solid phases at a given temperature in a closed system. At standard atmospheric pressure (1atm [standard atmosphere] or 101.3kPa [kilopascal]), water boils at 100°C. In other words, as the temperature of the water rises, the kinetic energy of the water molecules increases, and more of those molecules pass into the vapor phase, the water begins to boil. A liquid's boiling point is the temperature at which its vapor pressure equals the atmospheric pressure, so when that surrounding pressure is lower—as in high altitude conditions or under vacuum—the temperature at which the liquid boils is lower, too. Gale's reasoning behind pulling a vacuum on his brewing system checks out.

However, Gale's Frankenstein-esque coffee-making contraption, which essentially amounts to a vacuum reflux/distillation setup that's similar to a Florence siphon, is an overcomplicated mess that probably doesn't even function the way it's intended.

A proper siphon setup uses a boiling flask connected via a tube—outfitted with a filter that only allows liquid to flow through; this will make more sense in a moment—to a vessel that holds the waiting coffee grounds. A heating element boils the water in the flask, which then flows up through the tube and down into the collection vessel where it extracts the desirable compounds from the grounds. Once all of the water boils off and makes its way to the collection vessel, the source of heat is then removed from the boiling flask. The flask cools, creating a vacuum that actually pulls the brewed coffee from the collection vessel back through the tube to the original flask. That's where the filter comes in handy, to keep the coffee grounds out of your finished brew! (Note: The vacuum used to siphon off the coffee shouldn't be confused with Gale's idea of using a vacuum to boil the brewing water at a lower temperature to begin with. These are two separate uses of a vacuum for two distinct purposes.) Voila! Perfect coffee through physics!

In contrast, here's how Gale's coffee-making contraption is set up: An autoclave (a device that uses high temperature and pressure to sterilize equipment or solutions) is connected to a vacuum pump, which is then connected to a condenser. The condenser is nested in an Erlenmeyer flask that's sitting on a hot plate. The top of the condenser is connected to a T-connector that presumably collects the distillate and then drips it into a metal cylinder. This vessel appears to hold Gale's Sumatran coffee grounds since the tube that runs out of the bottom of it contains the brewed coffee. That tube then runs into a Florence flask, the output of which flows through yet another metal cylinder (the purpose of which escapes me). The final brew is then collected into a large vessel that looks like something like a burette (which is usually graduated since it is used for delivering measurable quantities of a liquid during titration; this one is not). If you didn't follow all of that, don't worry. This all looks very impressive when put together on screen, but needless to say, it's mostly nonsense.

Basically the first steps in this overall setup form an oddly redundant two-boiler system. First, the water is boiled under vacuum in the autoclave before it's collected in the condenser and boiled off again in the Erlenmeyer flask. While you could conceivably use an autoclave as a boiling kettle, that's definitely not the intended purpose for the vessel. And if you were boiling the water under vacuum to decrease its boiling point, you'd lose that effect by reboiling the newly condensed water in the Erlenmeyer flask, which is at atmospheric pressure.

The next thing that's incorrect about this setup is the connection to the condenser itself. The purpose of this piece of lab equipment is to expose vapors in the column to a relatively cooler surface—provided by cold water or air circulating in the jacket around it—in order to get the vapor to change phase into a liquid by, well, *condensing*. The condenser's input and output connections are for a circulating supply of cold water, not the brewed coffee itself. So this arrangement is unnecessary, but it looks cool.

The next step is a "black box" when it comes to the actual method of extraction: Is the metal cylinder a gravity-fed drip system or some sort of overwrought French press? I'm guessing the former since the distilled vapor is supposed to literally drip into the container and I don't see any mechanism for separating the extracted coffee from the grounds.

Then, the next series of question marks that occur about this head-scratching device is the connection to the Florence flask, a bit of glassware used for holding, boiling, and mixing liquids. In the setup, it just kind of sits there, presumably collecting the coffee ... which would *also* just sit there since there's no heat source to distill it (not that you'd want to) and there's no mechanical device to pump it over to the collection vessel. There *is* another nondescript metal cylinder between the flask and the final collection vessel, but if it's supposed to be a pump, it's in the wrong location; pumps work by *pushing* fluids, not pulling or "sucking" them out of containers. If that cylinder is supposed to draw the brewed coffee through it using the same phenomenon as the Florence siphon, it probably still won't work since there's no empty, cooling vessel—which creates the vacuum—to draw the coffee over. However, you can build a Florence siphon coffee maker of your own at home in order to get the perfect brew and one-up poor Gale Boetticher with your superior coffee-brewing knowledge.

This certainly isn't the last time that coffee makes an appearance on *Breaking Bad*, but it's the only time it's discussed so scientifically. I only wish that beer, the most ancient of libations, was revered on the show as much as its caffeinated counterpart. Luckily, Hank Schrader has so much respect for the potent potable that he makes time to brew his own beer in his man-cave, when not busting drug dealers and cartel members.

Biggest donation gets a six-pack of my very own Schraderbräu. Home-brewed to silky perfection.—Hank Schrader, Season 2, Episode 13: "ABQ"

I'll do my best to keep the discussion on brewing reined in a bit here, but it's easy to wander far afield and get lost in the wheat and barley. And while brewing is easy (and legal) enough to do at home, mastering the process takes a significant understanding of all the scientific disciplines that occur on a beer's journey from grass to glass. I'm talking botany, biology, chemistry, biochemistry, physics, microbiology, and engineering, all brewed into a perfect pint.

DEA Agent Schrader doesn't need an understanding of all of these areas of study just to make his home-brewed Schraderbräu, but a little more knowledge and attention paid to the process might have prevented his batch of beer from turning into "bottle bombs," a colloquial expression for beer bottles that quite literally explode. This is a real phenomenon that has a number of possible causes, which I'll explore shortly, but it's one that *Breaking Bad* gets right.

Hank's home-brewing hobby is an interesting and clever extra layer for his character. He spends his days (and most nights) tracking down drug dealers on behalf of the DEA, a stressful, adrenaline-fueled career that often puts him in the crosshairs of some very dangerous people. But even when he's not caught up in a desert gun battle or dodging the exploding heads of cartel members, Hank's dealing with illegal meth production and the physical and societal damage it causes on a daily basis. So it's interesting that his outlet—his own pressure-release valve, if you will—is a hobby that was once outlawed by the U.S. federal government. I like to think of Hank's home brewing as his own little rebellion (that and Cuban cigars), legal though it may be. His home-based brewing setup is even lovingly portrayed by the show's camerawork, which mirrors the showcasing of Walt and Jesse's RV lab, Gus's Superlab, and even Gale's coffeemaker.

But where Walt is a meticulous chemist who never seems to make a mistake when it comes to his cooks, Hank is a distracted and distraught home brewer when we see him in Season 2, Episode 5: "Breakage." Having just been promoted to Assistant Special Agent in Charge (ASAC) of the Albuquerque office of the DEA, Hank opts to take the day off, hoping to put a stop to his panic attacks by decompressing with a round of home brewing. Unfortunately, things soon go horribly wrong. Hank is awakened in the middle of the night by breaking glass and what sounds like gunshots. Expecting a home invasion, Hank draws his gun and investigates the disturbance only to find that his own Schraderbräu is the culprit, the bottles having exploded in the night.

So what's going on here? As I mentioned, "bottle bombs" are certainly something that can happen with beer, home-brewed or otherwise, but they're not common occurrences. Industrial brewing processes are designed to minimize breakage, both the literal kind and the general umbrella term that encompasses waste and profit loss. Most home brewers learn their lesson early on, either from their own "bottle bomb" experience or from their fellow brewers. There are myriad ways the brewing process can go wrong, but there are also literally thousands of years' worth of brewing experience to draw from to make sure it's done right. Even Hank admits to having been brewing for at least two years when he reminds his wife Marie of his own batch from Christmas 2006, so clearly he's been at this a while.

From the little evidence given in "Breakage," the bottle bombs have three likely explanations: contamination of the beer causing overcarbonation, structural failures in the glass bottles themselves, or overcarbonation during the brewing process. I'll talk about these in the order listed since they progress from least- to most-likely causes of Hank's bottle bombs.

If you walk away from this book with nothing other than the knowledge I'm about to impart, dear reader, then I will count myself successful. When it comes to a discussion of foreign objects in a brew, far too many home brewers refer to this as an "infection"; this is wrong. An infection is the invasion of an organism's tissues by disease-causing agents. Contamination is the presence of an unwanted constituent or impurity introduced to a material or environment. A stalk of grain, a single yeast cell, and the human body can be *infected* since these are living organisms susceptible to disease;

coffee, beer, and even meth can be *contaminated* since these are materials designed to incorporate desirable components while avoiding others.

Now that I've got that out of my system, let's talk about beer contamination. It can occur literally anywhere in the brewing process, from the growing grains in the farmer's field to the glass you pour your beer into. Contamination is an unfortunate certainty, but the good news is that there are plenty of steps along the way to minimize it and there are certain levels of acceptable contamination that won't ruin the finished product.

As for Hank's particular problem, there are a couple of culprits when it comes to contamination. It's possible that his bottles had some sort of contaminant in them—perhaps from an insufficient bottle cleaning—that caused overcarbonation either through chemical or physical means. In a beer bottle, even after the bulk of the brewing process is complete, yeast continues to metabolize existing sugars to produce flavor compounds, alcohol, and carbon dioxide in the beer. This gas builds up in both the liquid and the vapor within the bottle itself, providing the fantastic sound that can be heard upon popping the bottle's top and the wonderful fizzy sensation experienced while drinking. However, contaminants in the bottle or the glass itself can provide imperfect surfaces—called nucleation sites—for excess bubbles of carbon dioxide to form, possibly increasing the internal pressure to bottle-breaking levels.

An infamous wheat and barley fungus named *Fusarium graminearum* infects (used correctly here) the grain, causing blight and producing a mycotoxin called deoxynivalenol (a.k.a. vomitoxin). While this is a big problem for the livestock industry, when it comes to brewing, these particles—known more generally as hydrophobins, named for their "water-fearing" nature—are responsible for producing an effect that's more annoying than dangerous: excessive beer foam, or "gushing."[6] This tends to become very apparent once you actually open the bottle, but doesn't often cause bottles to break. So while it's possible that Hank got a bad batch of grain containing mycotoxins that contaminated (used correctly here) his brew with explosive consequences, there are more likely explanations.

The next most likely cause has to do with the structure of the beer bottles themselves. Home brewers are notorious for reusing bottles from brew to brew, which helps to cut down on costs but also opens the door for more contamination. However, reusing bottles or simply treating them roughly can introduce structural failures within them. Beer bottles are surprisingly highly engineered little works of art designed with internal pressures, ease of manufacturing, resistance to varying wavelengths of light, appearance, durability, and integrity in mind, among a host of other variables. Bottles made of green or clear glass allow more UV light to pass through than do brown glass bottles, which in turn allows light-sensitive compounds in hops to produce "skunky" aromas. Bottles of varying thickness can handle brews of varying carbonation, with thicker bottles used for higher-gravity beer and bottle conditioning. (The next time you happen upon a champagne bottle, take notice of how thick and heavy it is before popping the cork in celebration. You'll better appreciate the engineering that goes into a bottle holding back roughly three times the pressure of a car tire.)

Defects in these bottles can throw a wrench into all that engineering. Some defects occur during the manufacturing process by introducing cracks, chips, scratches, and bubbles into the glass itself. Other similar defects can occur during production, shipping, handling, and, yes, even the home-brewing process. Hank was pretty agitated while capping his latest brew of Schraderbräu thanks in part to his work-related panic attacks and the well-meaning concern from his wife, Marie. We even saw Hank shatter the neck of a bottle while cranking down way too hard on his bottle capper. It's possible that Hank got a bad batch of bottles that already had defects and were primed to fail, or that his existing bottles were damaged somewhere in his home-brewing process. However, the fact that quite a few bottle bombs went off suggests that something more systemic was at play rather than the variability of the bottles' structural integrity.

I'm guessing that Hank, experienced home brewer that he is, managed to overcarbonate his beer. This would explain why it wasn't just one bottle that went off, but nearly the whole batch. Sure, there's a chance that Hank's overenthusiastic capping introduced undue stress to the bottles, and there's even the possibility that a perfect storm of contamination, structural defects, and overcarbonation came together to trigger the bottle bombs. But we can explain most of this as simple overpriming.

Priming is one of the last steps in the home-brewing process. After you've gathered your grain and hops, steeped the grains, boiled the resulting liquid (called wort) along with the addition of hops, cooled the wort and transferred it to a fermentation vessel, aerated the wort and added the yeast, and allowed the fermentation to proceed for a couple of weeks, it's finally time to bottle your beer. Clearly, there are a lot of steps to consider and plenty of places for things to go wrong, but I'm going to pin Hank's troubles on this next step: the all-important priming step.

Priming sugar, used to produce carbonation in the bottle by allowing any active remaining yeast to gobble up the sugar and produce carbon dioxide and trace alcohols, is normally dissolved in boiling water and added to the beer before bottling. This step requires both proper timing and proper measurement; a mistake in either case could explain the overcarbonated bottle bombs. If Hank tried to bottle his Schraderbräu before his primary fermentation was complete, adding *more* sugar to the mix would have produced more carbon dioxide than he intended in *all* of the bottles, leading to their catastrophic failure in the middle of the night. Similarly, if he overprimed by exceeding the recommended amount (0.4oz to 1.1oz of priming/white sugar per gallon of beer) without using bottles engineered to withstand greater internal pressures, he was setting himself up for disaster.[7]

On the show, a few bottles exploded but the only real threat was to Hank's peace of mind. However, there have been anecdotal stories of the much larger half-gallon and gallon-sized growlers exploding due to high internal pressure, as well as the occasional freak accident that causes a keg to explode (though the latter has nothing to do with the relatively low pressure of beer under normal circumstances). Clearly the pressure building within the bottles was meant to represent the pressure building within Hank himself, but luckily, he proved to be a more durable vessel than his bottles of beer, at least for another couple of seasons.

A note on just what kind of beer Schraderbräu actually is: Some folks have surmised that it's a lager while others have taken issue with this idea since Hank conditions his bottles on the countertop and not in a refrigerator. Lager yeast ferments more slowly and at colder temperatures than ale yeast, say around 40–50°F, so these brews need cold conditioning or "cold storage" to finish properly. (Lagers get their name from the German word *lagern* meaning "storage.") But Hank was likely brewing his batch in late November 2008—at least according to the unofficial *Breaking Bad* timeline—in a garage (without climate control, presumably) in Albuquerque, where the average temperature was 47°F. That temperature is just fine for lagering.

However, as I discussed earlier, Hank's bottle bombs likely resulted from overcarbonation, a problem that's far more likely in an ale than a lager due to the faster fermentation rate and warmer fermentation temperatures. Since Hank styled his beer with the German-sounding name Schraderbräu, I'm going to guess that he's brewing up a classic German Kölsch. (And, no, I can't explain Hank's Hawaiian shirt, lei, or "hang ten" hand sign in the logo, but the German beer stein makes sense, at least.) The Kölsch is a good candidate because it's warm-fermented using ale yeast (allowing for a more likely overcarbonation explanation), but it's also cold-conditioned, which can be explained by his countertop conditioning in the Albuquerque winter. Hank's home-brew recipe has proven to be even more mysterious than Heisenberg's own cooking process, so we may never know. Either way, the next time you find yourself drinking a freshly brewed pint, remember to pour one out for ASAC Schrader.

Inside *Breaking Bad*: As writer Moira Walley-Beckett revealed in the *Breaking Bad Insider Podcast,* the "Schraderbräu" song that Hank can be heard singing is actually a take on the old Löwenbräu beer jingle; music supervisor Thomas Golubic secured the rights to the song for that episode. Swedish director Johan Renck also taught Dean Norris a 200-year-old Swedish drinking song as a backup in case they couldn't get the rights. They used both songs for Season 2, Episode 5: "Breakage."[8]

Side RxN #6: They're Minerals, Marie!

After Hank's home-brewing hobby went bust thanks to his panic attacks, he took up the relatively safer and, subjectively, more mundane pastime of collecting rocks. (I'm sorry, *minerals*.) Hank's new obsession with mineralogy came after barely surviving the shootout with "The Cousins" (Leonel and Marco Salamanca, twin brothers and hitmen for the Juárez cartel); the temporary paralysis of his legs left him bed-ridden with lots of free time on his hands. Actor Dean Norris wasn't totally sure what the hobby meant to Hank, but suggested, "Cataloging those minerals kind of gave him some order to the chaos of his life."[9]

Minerals are, by definition, naturally occurring chemical compounds, each with one specific chemical composition in which its atoms and molecules are highly ordered, forming a crystal. There are around 5,300 known minerals, giving Hank plenty of things to study and add to his collection, much to his wife's displeasure. (He did, however, use fake mineral conventions as cover for his investigative work later on.) Rocks, on the other hand, can be made up of multiple minerals.

Specific minerals that Hank is seen studying include a sample of magnesite, a.k.a. magnesium carbonate ($MgCO_3$)—which ranges in color from colorless to pale yellow to lilac-rose—and rhodonite ($MnSiO_3$). In Season 4, Episode 4: "Bullet Points," Hank correctly identifies the latter as a "manganese inosilicate," a fancy way of saying the mineral is composed of manganese and interlocking chains of silicate (SiO_3) groups. He also explains to Walt Jr. that when manganese oxidizes, like rust, it turns pink. Master chemist Walt steps on Hank's budding mineralogist toes and rattles off manganese's oxidation states "between –3 and +7, which takes it through a range of colors," including "its most stable-state +2, which is pale pink." Walt may be right, but his conversation and room-reading skills need a polish.

You can draw any number of inferences from Hank's study of crystalized minerals as related to his investigation of crystal meth; ultimately, it was a minor plot point. Hank may have been happier taking down Heisenberg, but it's clear he would have been better off sticking with the noble pursuit of mastering mineralogy.

Chemistry II

Welcome back to the chemistry section of *The Science of "Breaking Bad"*! Now that I've laid down the fundamentals of elements and their particles, and explained how chemistry can quickly turn deadly in the wrong hands, it's time to get a little more advanced. Walt's superior knowledge of chemistry served him well in the classroom, but it was in the criminal world that it became an essential skill in Heisenberg's arsenal. Sure, that knowledge mainly supplied him with an endless stream of methamphetamine and ever-increasing stacks of cash, an important plot point in *Breaking Bad*. But just as chemistry allows for the creation of all-new compounds, so, too, does it offer the power to destroy.

In this section, I'll take a look at the more destructive side of Heisenberg's mad science. I'll revisit the fan-favorite explosives—fulminated mercury and the wheelchair bomb—to see if the on-screen chemistry lives up to its potential. I'll then pick apart his scheme to cobble together an industrial-strength DIY lockpick through the use of thermite. And finally, I'll cover the basics of acids in a breakdown of one of the show's most frequently used chemicals: hydrofluoric acid. Things are about to get a little messy.

VIII Explosives:

Fulminated Mercury and the Wheelchair Bomb

From Dr. Donna J. Nelson:

There was initially a wide divergence of attitudes about *Breaking Bad.* This changed during its run from inception to finale.

During Season 1, few scientists seemed to be aware of the show's existence. By Season 2 that had changed, and it was fairly obvious that it would be successful. A look at the Rotten Tomatoes ratings for the show's five seasons quantifies the show's popularity among the public; by Season 2, the rating had reached 100, and it remained there for the rest of its seasons.

Consciousness of the show among scientists and especially chemists increased as a result of the 2008 article about it in *Chemical & Engineering News.* Most scientists who were aware of the show appreciated its science and liked it generally otherwise.

To some extent, when scientists possessed extreme opinions of the show, the attitudes were age-dependent. The few negative remarks I received about my participation originated almost exclusively from older scientists. A couple of chemists expressed concerns at ACS National Meetings, and two letters were mailed to my work address. They all worried about the show tarnishing the good reputations of scientists with the public. However, they didn't understand that much of the perceptions of science and scientists held by the public have come from Hollywood, and in order to have a chance of changing the way Hollywood portrayed us, we scientists would need to work with them, and we were not in a position to choose or dictate what form that would take.

In contrast, students loved the show from start to finish, especially students in scientific disciplines. At my home university, students asked to come chat with me or asked if I would give an opinion regarding a project for a class or a degree. These were not merely students in science and engineering programs, but students enrolled in business and education. I didn't understand how the opinion of an organic chemist would pertain to a business proposal, but the students emphatically claimed it did. And they wanted to take selfies. I accommodated all I could.

The students who amazed me the most were in high school, turned onto science by the show, and eager to start their own science blog. They emailed me, asking me to answer a few questions, which they could post in order to get their blog started. I always helped them.

101

Some of *Breaking Bad*'s most memorable moments are literally the show's most explosive. On two separate occasions, Walt uses his knowledge of chemistry to create explosive compounds in order to gain leverage: One is used as a bargaining chip with small-time drug lord Tuco Salamanca, and the other is used to dispatch the intimidating drug emperor Gus Fring. But to fully appreciate the powerful potential of explosives, we need to understand their chemistry.

Explosives have been around a long time, since at least the Chinese invention of black powder used for mining and military purposes in the ninth century. An explosive can be defined as a reactive substance that contains a great amount of potential energy that can be released suddenly in an impressive display of light, heat, sound, and pressure, otherwise known as an explosion. These explosives have less potential energy than fuels, but the release of this energy happens at a much faster rate. In *chemical* explosives, this energy is stored as *chemical* energy in the bonds between atoms.

As Walt mentions to his class in Season 1, Episode 6: "Crazy Handful of Nothin'," chemical reactions involve change on two levels: matter and energy. (Keep that theme of "change" in mind throughout, whether we're talking about making meth, explosive reactions, or Walt's metamorphosis into Heisenberg.) The relationship of heat and temperature to energy and work is best explained through the physics branch of thermodynamics, but we'll focus our discussion on Walt's chemical concoctions in this section. A gradual reaction can proceed slowly or result in a barely noticeable change in energy, while an *explosive* reaction has quite the opposite result. In chemistry terms, in an explosive reaction, the change from reactants to the more stable products is accompanied by a great release of heat, gas, and much sound and fury. And when a chemical explosive is as highly sensitive as fulminated mercury—due to its sensitivity, the compound is stored under water to prevent self-detonation until it's ready to be used[1]—it doesn't take much to trigger the explosive reaction and release all that stored-up energy.

In this same episode, Walt also delivers some excellent foreshadowing disguised as a classroom lesson on explosive chemicals, like the aforementioned fulminated mercury. This exotic-sounding material is an example of a primary explosive, a chemical compound that can be detonated without much external stimulation through impact, heat, or electricity, to name a few methods. Later in the episode, Walt actually has this particular explosive chemical compound on hand in order to keep drug dealer Tuco Salamanca and his goons at bay. Walt uses a relatively small amount (50g) to make his point, detonating it by throwing it on the ground like an amped-up bang snap, one of those novelty noisemakers that make a little explosive pop when thrown down. Walt's detonation, however, results in a concussive blast that blows out the windows of Tuco's headquarters.

But while fulminated mercury is a very sensitive primary explosive, *Breaking Bad*'s depiction of it does not line up with the compound's real-world appearance, potency, or sensitivity. Adding insult to literal injury, a blast of the size seen in the episode would definitely have injured everyone in the room and would also have exposed them to toxic mercury dust to boot. Sorry, Heisenberg. (A *MythBusters* special did a fantastic job of breaking this scene down if you're interested in seeing how it all plays out after learning the science behind the scenes.[2])

But Walt's explosive experimentation doesn't end there. In one of the most shocking moments from the entire run of *Breaking Bad*, Walt takes out two of his antagonists in one fell swoop by pitting their mutual animosity against each other. This time, Walt isn't aiming for intimidation or distraction but rather, outright death and destruction.

In Season 4, Episode 12: "End Times," Walt can be seen harvesting cold packs for their potentially explosive chemicals and combining them with common vegetable oil, which apparently results in a bubbly, charcoal-colored sludge when boiled. We next see a metal canister laid out on the kitchen table while Walt tinkers with a battery-powered circuit board that produces a spark when he presses a button on a walkie-talkie. What's all this mad science about? A homemade pipe bomb, of course. Walt's small-scale test in the kitchen is enough to trigger a tiny amount of detonating compound, but the big boom comes in the Season 4 finale, "Face-Off."

Essentially, Walt makes a DIY ANFO bomb. ANFO is a precisely mixed combination of compounds (more on this mixture later) that makes for a cheap and reliable explosive widely used in a variety of industries, though the chemistry is sometimes co-opted for malicious purposes—**again, not the purpose of this book**. Putting that explosive inside a metal canister to turn it into a pipe bomb and then strapping it to an old man's wheelchair in a nursing home is about as malicious as it gets. (The writers were careful to note in the radio news report following the blast that "only" three people were reportedly killed—Gus, his enforcer Tyrus Kitt, and former high-ranking cartel member Hector Salamanca—avoiding any collateral damage.)

We can presume that master-chemist Walter White would have no problem making a batch of fulminated mercury or figuring out the chemical composition and mixture of an ANFO bomb, but to see just what made his chemical concoctions so deadly, we'll have to get a little more advanced.

Advanced

Fulminated Mercury

The faster reactants, i.e. explosives—and fulminate of mercury is a prime example of that—the faster they undergo change ... the more violent the explosion.—Walter White, Season 1, Episode 6: "Crazy Handful of Nothin'"

Let's start with the chemistry of fulminated mercury, also known as mercury (II) fulminate, or $Hg(CNO)_2$. Mercury should be familiar to anyone who's used an old-school thermometer, so it's more likely that the word "fulminate" is new to you. It hails from the Latin *fulmen*, meaning "lightning." As a verb, it means to either explode or detonate. As a chemistry term, it specifically refers to an explosive salt containing the CNO^- group. This group is thermodynamically unstable due to its structure, which places undesirable formal charges—the charges assigned to an atom assuming an equal sharing of electrons in chemical bonds—on each atom of the CNO group, with mercury (Hg) as the central atom.[3]

In other words, it's an unhappy little molecule that will do just about anything to get to a more stable state, even if that means releasing a bunch of explosive energy with just the slightest encouragement. The possible products of the explosive reaction include elemental mercury, carbon monoxide, carbon dioxide, and nitrogen, which are much more thermodynamically stable than mercury fulminate itself.

Fulminated mercury can be prepared by a combination of mercury, nitric acid (HNO_3), and ethanol (C_2H_6O), as was discovered by Edward Howard in 1800.[4] The resulting whitish powder—not crystals like Walt's meth mimic—can be used as a primary explosive; it was used in older models of percussion caps and blasting caps to trigger other more powerful but less sensitive explosives, known as secondary explosives. Mercury fulminate's sensitivity to friction, shock, spark, and heat makes it ideal for such a purpose.

The chemistry of mercury fulminate's preparation has been around long enough that Walt should have had no trouble cooking it up on the side, though *Breaking Bad* creator Vince Gilligan suggested in the *MythBusters* episode that Walt may have also mixed in the even less stable silver fulminate to get a bigger bang for his buck.[5] This relatively more explosive compound has limited uses since it's so sensitive and can't be aggregated in great quantities without self-detonating under its own weight. However, silver fulminate is used to coat the gravel in "popping" noisemakers or bang snaps, as well as Christmas crackers, though in such small quantities of the explosive as to render the novelties harmless. Walt's "tweak of chemistry" was anything but.

In order to detonate his sample of mercury fulminate disguised as crystal meth, Walt throws the explosive onto the floor of Tuco's headquarters as hard as he can. Unfortunately, as the *MythBusters* team members themselves also showed, Walt would have had to throw the explosive down with superhuman force since even their amped-up robotic throwing arm could not get the real thing to detonate on impact.[6] This is why modern detonators and blasting caps use an electric spark as a more reliable trigger, despite the fact that Edward Howard found more than two centuries ago that when striking "3 or 4 grains" of the stuff with a hammer, "a very stunning disagreeable noise was produced, and the faces both of the hammer and the anvil were much indented."[7]

That charmingly antiquated description brings us to the next scientific aspect of this scene: fulminated mercury's explosive potential. In the episode, the detonation blows out the windows of Tuco's headquarters, knocks an air-conditioning unit from its perch, and leaves Walt and the gangsters standing in a cloud of dust. Though Walt used 50g of the crystallized powder—far more than Howard's three or four grains and resulting in much more than "a very stunning disagreeable noise"—the *MythBusters* once again busted this bit of Hollywood trickery.[8]

While 50g of fulminated mercury certainly produced a concussive blast in their test, it failed to blow out the windows of their mock-up scene; 250g, on the other hand, decimated the makeshift building completely. Perhaps Gilligan's earlier assertion of additional silver fulminate in the mix could have made up the difference in explosive potential, but ultimately, it's a moot point. Why? Because Walt and the gang members would not have walked away from that explosion unharmed.

Though the concussive blast from 50g of mercury fulminate was not enough to blow the windows out of Tuco's place, it *was* enough to injure the bystanders, as the *MythBusters* discovered.[9] In other words, even if Walt's 50g sample packed some extra oomph thanks to silver fulminate, he would have borne the brunt of the blast. That's not how it plays out in the episode since Walt survives unscathed to strike a deal with Tuco and walk away with both his bag of explosives and $50,000 in cash. But even if the drug dealers somehow managed to survive the blast, there are other inherent dangers when dealing with this particular combustible.

In addition to its explosive properties, a mercury fulminate explosion can also emit toxic nitrogen oxide fumes and aerosolized mercury salts upon combustion. So even if Walt's homemade bomb hadn't blinded the drug dealers with science, he could at least have played the long game and threatened them with long-term respiratory disease and mercury poisoning. But, of course, that ending isn't nearly as explosive. And that's where Hollywood comes in again.

Inside *Breaking Bad*: Gilligan and his writers knew that Heisenberg would not be able to beat Tuco's gang through strength of arms or some sort of shoot-'em-up, so they relied on his supervillain skill: science! Clearly, they didn't use the actual explosive in question for the scene. Instead, the show's special effects techs used PRIMACORD, a brand of detonating cord or "explosive rope," to pull off the explosion once cast and crew were removed to a safe distance. They actually reined in the size of the explosion seen on camera to make the survivability of the characters more believable.

In fact, Bryan Cranston didn't even throw anything at the floor at all. A combination of a special camera rig and a bit of prop trickery was used to get this slow-motion shot. The rig uses a platform to raise the actors above the camera and also places the camera lens itself behind a transparent piece of glass or plexiglass; the resulting effect can be seen throughout the series whenever you're looking up at a character from a low angle and through an otherwise opaque surface. If you watch the scene again, you can see Walt turn to throw the fulminated mercury at the ground, and then, after a cut, the explosive comes directly at the camera before bursting with a bright flash and concussive force. That effect was actually achieved by using a thin dowel with the fulminated mercury packet held on the end of it, which was pushed toward the camera lens. No explosives here, just a little bit of misdirection.[10] It's a fine example of the difficult balance between cinematically exciting moments and real-world science, a balance struck quite well by *Breaking Bad*.

Wheelchair Bomb

> Skyler White: Jesus, Walt, the news here. Gus Fring is dead. He was blown up along with some person from some Mexican cartel, and the DEA has no idea what to make of it. Do you know about this? Walt? I need you to—
> Walter White: It's over. We're safe.
> Skyler White: Was this you? What happened?
> Walter White: I won.—Skyler and Walter White, Season 4, Episode 13: "Face-Off"

Like many other sections of this book, the following explanation of ANFO and its explosive potential is for scientific understanding *only* and without intention or encouragement of DIY experimentation. As Gilligan himself said in a related episode of the *Breaking Bad Insider Podcast*, "Knowledge is knowledge, and it's either good or bad depending on what you [do] with it."[11] The reasons behind this disclaimer should be obvious, but it's also worth mentioning that Walt's DIY explosive was of a similar concoction to that of the domestic terrorist truck bombing of the Alfred P. Murrah Federal Building in Oklahoma City in 1995, which claimed the lives of 168 people, injuring 680 others.

Now back to the science! As I mentioned earlier and as Gilligan himself confirmed in the podcast, Walt's DIY project in "End Times" was a homemade pipe bomb powered by an explosive known as ANFO.[12] The chemical mixture gets its name from its components: ammonium nitrate (NH_4NO_3) or "AN" and fuel oil or "FO." That's it. It's relatively simple to make and, though the mixture needs to be precise with a majority of ammonium nitrate relative to fuel oil, it's also quite cheap. It's so simple and cheap, in fact, that it's favored in mining operations, quarrying, certain civil construction projects, and even avalanche hazard prevention.

Now let's break ANFO down into its components. Ammonium nitrate, a nitrate salt of the ammonium cation, has quite the disparity in its uses: agriculturally speaking, it's used as a high-nitrogen, single-nutrient fertilizer since it takes plants a long time to acquire nitrogen naturally; industrially it's used as a component of the explosive ANFO, as I've mentioned. Ammonium nitrate by itself is nonexplosive, however. While the mineral occurs naturally and can be mined, most of it is produced synthetically owing to the vast quantities required for both agricultural and industrial applications.

So what is Walt doing cutting open a bunch of cold packs? The answer: a clever use of ammonium nitrate, which can be found in these particular types of portable "instant" cooling devices. Essentially, each cold pack consists of two bags: one that's full of water inside another bag containing ammonium nitrate or similar chemicals. These cold packs will stay at ambient temperature until squeezed, which breaks the water bag and dissolves the solid ammonium nitrate, producing an overall endothermic reaction, which I will detail.

Ammonium nitrate's dissolution in water actually requires energy—in the form of heat—pulled from its surroundings, the water itself. The ammonium nitrate is composed of ammonium ions (NH_4^+) and nitrate ions (NO_3^-) held together in a crystal lattice by ionic bonds, just like table salt, NaCl (Na^+ and Cl^-). Calculating just how much energy is needed or released depends on the sum of three enthalpy values: the breaking of the bonds between the solute's molecules forming the crystal lattice structure, the breaking of the bonds between solvent molecules (like hydrogen bonds in water), and the forming of new solute-solvent bonds to fully dissolve a solid. For both NH_4NO_3 and NaCl, the first two steps require more energy to break bonds (endothermic) than is released by the forming of new bonds (exothermic), but dissolution of ammonium nitrate is roughly seven times more endothermic than table salt under ideal conditions.

Put more simply, the introduction of water starts to dissolve the NH_4NO_3, which requires heat energy absorbed from the surrounding water itself. This has an overall cooling effect, making the portable packs useful for treating minor injuries, or if you're Heisenberg, for harvesting ammonium nitrate. (Sure, Walt could have purchased a bit of fertilizer, though the U.S. Department of Homeland Security has proposed regulation through registration for such sales.[13] He probably could have synthesized it himself, but that would have been far less interesting to shoot.)

The other ingredient in ANFO, fuel oil, is a liquid fuel that's burned in a furnace or boiler for heat, or used in an engine to generate power. Types of liquid fuel include kerosene or paraffin, and diesel. And, both in the real world and the world of *Breaking Bad*, since simple vegetable oil can be used as an alternative fuel in diesel engines and heating oil burners, it can conceivably act as a fuel oil substitute. It's a quick shot, but you can clearly see a bottle of opened vegetable oil on Walt's kitchen counter during his ANFO manufacturing, which is a clever inclusion.

But how do such relatively benign chemicals combine to form an explosive that's capable of killing people? The specific chemistry of the ANFO reaction depends on the type of fuel oil used, but the products are nitrogen, carbon dioxide, and water, along with some toxic byproducts like carbon monoxide and nitrogen oxides. When a reaction produces nitrogen (N_2), as is the case with both fulminated mercury and ANFO reactions, watch out. To produce the incredibly stable diatomic gas from less-stable nitrogen reactants, a *lot* of energy is released, which is why nitrogen-producing reactions tend to be very exothermic, rapid, and explosive. You only have to look at the devastation caused by terrorist bombings or fertilizer plant explosions to get an idea of just how powerful nitrogen reactions can be.

Classified as a blasting agent, ANFO needs to be set off by detonation of a high explosive; contrast this with fulminated mercury's status as a primary explosive *used* as a detonator. ANFO, a tertiary explosive, requires a secondary explosive like dynamite for detonation since it's so insensitive to shock. Picture a black powder or PRIMACORD fuse that detonates a primary explosive blasting cap, which detonates a secondary high-explosive booster like dynamite, which then detonates the ANFO. That complicates Walt's pipe-bomb setup a bit, but I'll assume he took care of the necessary detonation chemistry off camera.

Once ANFO is detonated, things get messy quickly. First, a detonation wave is produced that travels outward at two to three miles per second through the surrounding ANFO mixture. This vaporizes the ammonium nitrate solids in an instant, breaking the molecules down and forming oxygen in the process. The oxygen combined with the energy of the detonation wave ignites the fuel, which rapidly combusts and produces even more gas. The quickness with which this gas is generated produces pressure waves that travel at the speed of sound. Substantial heat is produced, but it's actually these pressure waves that do most of the damage. Anyone who's ever watched the *MythBusters* team use ANFO to blow something up—like when they vaporized a full-size cement truck—knows how destructive an explosion's pressure wave can be. Gus, Tyrus, and Hector found this out the hard way.

However, some fans were upset at the apparent plot hole in Walt's plan in this episode. Their reasoning goes like this: Tyrus had checked Hector's room—and his wheelchair—ahead of time with both a visual scan and a sweep for radio transmitters, since he and Gus were taking precautions against any potential assassins and/or listening devices. So how did Tyrus miss the giant metal canister attached to the wheelchair? And if the bomb was actually remotely detonated, how did his scanner miss the radio signal-transmitting device?

I think this question can be answered by a combination of editing choices made in the episode and a closer look at the device in question. As far as the edits go, the final cut shows Walt in the nursing home room asking Hector if he has any second thoughts before presumably rigging the bomb. After this, Tyrus enters the room and does his sweep while Walt hides outside by the window. The confusion in the order of events continues since the edit then shows Walt driving away from the nursing home before Gus arrives for his final visit, presumably leaving Tyrus inside the room the entire time. This makes sense narratively, but doesn't explain why Tyrus missed the giant, obvious, crude pipe bomb attached to the wheelchair. (You could make a case for Tyrus mistaking this for part of the oxygen tank setup, but c'mon.) It would make more sense if Walt had waited outside until Tyrus had completed his sweep and then rigged the bomb once the henchman stepped outside for a coffee break. Both approaches stretch credibility just a tad and I'd imagine that the creative team had a similar discussion during editing; we'll chalk it up to suspension of disbelief.

As for the method of activation of the bomb itself, this is more concrete. Yes, Walt tested his homemade explosive in his kitchen by using a walkie-talkie as a remote detonator (which wasn't super reliable), but for the full pipe-bomb setup, you can clearly see that Hector's bell is hardwired to the explosive device. This gave Hector full control over the timing of the detonation and provided a more reliable trigger than the radio-activated device (also explaining why Tyrus's scan wouldn't pick up a transmitter); it also allowed Mark Margolis a final moment to show off his powerful acting abilities as the mute-yet-communicative Hector. The idea that Walt was somewhere nearby and triggered the bomb upon hearing the bell is understandable but incorrect, in my view.

Inside *Breaking Bad*: From the Hollywood side of things, this incredible scene was pulled off with a combination of visual effects work, computer-aided effects work, makeup, and clever editing. All of these departments worked together to bring about the stunning moment when Gus Fring walks out of the room post-explosion, initially appearing unscathed when he is seen in profile before the camera shifts to a frontal view to reveal that the other half of his head is now skeletal and empty. And it all had to be done safely, obviously, despite the destructive demonstration on display. To pull off Gus's gruesome transformation, Greg Nicotero and the fantastic effects team from AMC's *The Walking Dead* lent a hand (and a face) during many weeks and months of special makeup effects work. I don't think I'm alone in saying the hideous result was well worth the effort.[14]

Side RxN #7: Quality Face Time with Gus Fring

It's not all murder and meth on *Breaking Bad*; sometimes even the show's worst villains break bread instead. In "Abiquiu," the eleventh episode of Season 3, Gus invites Walt to dinner where he's serving a specially prepared Chilean dish. While they chow down, Gus remarks on his amazement over the connection between the senses and memory, to which Walt says, "Basically, it all takes place in the hippocampus. Neural connections are formed. The senses make the neurons express signals that go right back to the same part of the brain as before, where memory is stored. It's something called 'relational memory.'"

Walt may not be a neuroscientist, but he's got the basics down. "Relational memory" is a type of memory that relates different items of an experience, like associating a name with a face or remembering the order of events. It's different from both declarative memory—facts and events—and procedural memory—acquiring and expressing skills—to name but two types of memory.

There's a lot going on in the brain and its 100 trillion synapses, or structures through which one neuron can transmit an electrical or chemical signal to another neuron. In the case of relational memory, the process starts with sensory inputs, like certain sights, sounds, or smells. Those various sensory inputs are then analyzed by different regions of the cerebral cortex, which essentially sort out the "what" (the aroma and taste of Chilean fish stew) and the "where" (Gus's childhood home) into two different streams of information. When these streams eventually converge in the hippocampus, context is applied, and the streams are bound together. That information can then be consolidated from short-term to long-term memory storage in the hippocampus. These memories are not only strengthened through repetition—depending on how often Gus's mother made paila marina—but can also connect to other memories related through context.

Smell is a strong trigger for emotions and memories due in part to the sensory input's path through the olfactory bulb, which has direct connections to both the amygdala and the aforementioned hippocampus, just two parts of the emotion/memory-specific limbic system. Visual, auditory, and tactile inputs don't take this particular pathway, so they're not as firmly rooted in memory as olfactory inputs are.

IX Pyrotechnics:

Thermite Lockpick

From Dr. Donna J. Nelson:

When I gave invited talks on my assistance with the science in *Breaking Bad*, attendance frequently overran the room. The audience, mostly students, sat in the aisles and on steps, stood in the back, and filled overflow rooms. After the talks, I stayed for autographs and selfies, sometimes thirty to forty-five minutes. I was amazed at the size of the audience that my *Breaking Bad* work drew.

At one such invited talk, a person wearing a yellow hazmat suit—familiar Walter White attire!—brought a glass of wine to me at the podium. I kept speaking, and the audience loved it. This attendee was a visiting student from Germany, where the show was immensely popular. He was a member of the Young Chemists Group of the German Chemical Society, and he communicated to other members that I was a good sport and gave a good talk on a favorite show. Shortly afterward, I received an invitation from the Young Chemists Group to take a speaking tour through Germany.

Some of the largest audiences for my speaking engagements were during my trip through Germany in May 2014, for the Young Chemists Group of the German Chemical Society. They hosted me for a six-day speaking tour in six different cities: Frankfurt, Essen-Duisburg, Kiel, Dresden, Bayreuth, and Potsdam. At each location, some people attended the talk dressed in hazmat suits, like the one Walt is wearing in the poster shown here, which advertised my presentation at Essen-Duisburg.

I was told that in the more southern and eastern German cities on my tour, the audience might appear a little rougher; these were the cities where drug use was higher. The prediction proved true. The audience had a higher population of biker jackets, draped clothing chains, face piercings, and tattoos. I worried a little as I gave my regular talk, omitting nothing, including my comments warning against synthesizing and using illicit meth. I didn't want to insult or anger anyone. Nevertheless, after my talks in those cities, audience members still formed long lines to get my autograph and a photo with me.

Figure 9.1
Poster from Dr. Donna J. Nelson's presentation in Essen-Duisburg, Germany. Image courtesy of Dr. Donna J. Nelson.

101

In World War II, the Germans had an artillery piece—it's the biggest in the world—called the Gustav Gun, and it weighed a thousand tons. And the Gustav was capable of firing a seven-ton shell and hitting a target, accurately, twenty-three miles away. I mean, you could drop bombs on it every day for a month without ever disabling it. But, drop a commando—one man—with just a bag of this, and he could melt right through four inches of solid steel and destroy that gun forever.—Walter White, Season 1, Episode 7: "A No-Rough-Stuff-Type Deal"

When Walt and Jesse find themselves short on methylamine in the Season 1 finale, they are faced with two options: Pay $10,000 to professional thieves who will steal the precursor chemical from a guarded, secure warehouse, or steal it themselves. (One guess which option the money-minded Walt decides to take.) However, one big thing standing in their way—besides the obvious hazards of subduing guards and avoiding being identified by security cameras—is the heavy-duty lock on the steel door barring the entrance to the supply room. Luckily, Walt gets a bolt of inspiration from a simple children's toy: the Etch A Sketch.

Since *Breaking Bad* clearly didn't have the rights to the Ohio Art Company toy, now owned by Spin Master since 2016, the props department got creative with it. Walt picks up a green version of the familiar red drawing toy, one that reads "Sketcher-roo: It's fun! It's easy!" In keeping with the spirit of the show (and the legal department), I'll be referring to it as a Sketcher-roo from here on out. Jesse's confusion as Walt holds up this simplistic toy as an answer to their security problems is understandable, which is why I'm going to sort out this particular bit of *Breaking Bad* science right now.

You may never have wondered just how a Sketcher-roo makes the pictures that it does, but this is as good a place to learn that factoid as any. Inside the Sketcher-roo is a metallic powder that coats the inside surface of the glass display. It's this powder that is scraped away, using the knobs that control a stylus, in order to make a drawing. (There are also small plastic beads mixed in to keep the powder flowing freely, which is what accounts for the familiar sound you get when you turn the toy upside down and shake it to erase the drawing, depositing powder back onto the glass.) Walt knows that this particular powder is one of two components in a very potent mixture called thermite. By simply placing the bag of homemade thermite on top of the facility door's substantial lock and setting it on fire with a blowtorch, Walt and Jesse are able to break into the chemical supply storage with relative ease. Getting the methylamine out, however ... Well, as Hank says while reviewing the security footage, "Try rolling it, morons. It's a barrel. It rolls."

Common sense aside, Walt's lockpicking chemistry knowledge is a surefire recipe for success here. Some of the details found in his story about the "Gustav Gun" and the ease with which he gathers the metallic powder from a handful of Sketcher-roos don't quite add up, but for the most part, this one's solid. Thermite, when ignited, goes through a very energetic chemical reaction that gives off enough heat to melt or cut through metal, making it a go-to option for on-site welding; it's also been used in military applications to disable equipment and weaponry. This surprisingly simple compound, which even burns underwater, could certainly melt away a steel lock, but to find out exactly how that happens, our discussion's going to have to get a little more advanced.

Advanced

In simple terms, thermite is a composition of metal powder and a metal oxide. That's it! (Note: This is not a 1:1 mixture, but neither is this a DIY instruction manual, so **consider this a reminder not to try this stuff at home.**) When ignited, thermite gives off heat, light, sound, and gas or smoke in an exothermic redox reaction, making it a favorite display for classroom chemists and YouTube enthusiasts. However, it's *not* an explosive, like some of the compounds we discussed earlier. The nondetonative, self-sustaining reaction of thermite, known as a pyrotechnic composition, doesn't rely on external oxygen to sustain itself but rather the oxygen provided by the metal oxide. Other types of pyrotechnic compositions that might be more familiar include flash powder, gunpowder, and sparklers or flares.

The word "redox" is a portmanteau of reduction and oxidation, the two processes that occur in a reaction like the one thermite undergoes. The terms "reduction" and "oxidation" define what happens to certain reactants when electrons are transferred between them during a reaction: A reactant is reduced if it gains electrons, decreasing its oxidation number; a reactant is oxidized if it loses electrons, increasing its oxidation number. In the case of the thermite reaction, the iron is reduced and the aluminum is oxidized. Here's a look at the thermite equation including the oxidation numbers representing the number of electrons gained or lost:

$$\overset{[+3][-2]}{Fe_2O_3} + 2\,\overset{[0]}{Al} \rightarrow 2\,\overset{[0]}{Fe} + \overset{[+3][-2]}{Al_2O_3}$$

Redox reactions like the one involving thermite are quite common. They occur in the body's metabolic processes, the ubiquitous process of rusting, and in common, everyday batteries. I briefly touched on this chemistry in chapter 5, on Walt's DIY battery, but it bears a little more discussion. Each and every redox reaction includes two half-reactions, one for reduction and one for oxidation; these occur simultaneously and not independently of each other. And in the case of thermite, this is also a competition reaction: aluminum is a more reactive metal than iron (i.e., more willing to give up its electrons) so the former will displace the latter, as you can see in the preceding equation.

Discovered and patented in the 1890s by German chemist Hans Goldschmidt, who gave his name to the process, the reaction was originally intended to produce pure metals without using carbon in the smelting process. However, thermite soon became a valuable commodity for commercial and industrial welding.[1] Since it's not explosive and exposes a small area to very high temperatures—on the order of about 2300°C or 4200°F—over a short period of time, the reaction can be used to either cut through metal or weld metal components together. Thermite is often used to join two sections of railway track on site by allowing the molten material that's produced from the reaction to flow through a mold around the joint. Thermite can also be used for repairs (via welding), purifying ores of some metals (as was its initial intended use), or even welding thick copper wires together for electrical connections used in electrical utilities and telecommunications industries. Military uses of thermite include incendiary bombs, as a component in compounds meant to destroy equipment that might fall into enemy hands, or to disable artillery without using noisy, explosive charges. Since it's such a potent tool for melting and cutting through metal, there's every reason to believe Walt's homemade thermite would have cut right through the storage lock.

To be clear, the composition of thermite itself—the metal powder fuel and the metal oxide—is quite stable at room temperature, which is why Walt can toss it to Jesse without a care and why it's so suitable for commercial and industrial uses. However, this stability also makes thermite difficult to ignite in the first place. A cigarette lighter or match won't cut it. Since thermite requires a temperature above 3000°F to ignite, a magnesium ribbon "fuse" is ideal since it burns around 5610°F. Walt uses what appears to be a handheld propane torch to light his package of thermite, which might just have been able to get the job done at a maximum temperature of 3623°F. But once the thermite gets cooking, watch out, because you're not going to be able to stop it. (Forget using water; thermite is also used for underwater welding.)

Here's another look at the chemical equation of the thermite reaction using iron (III) oxide and aluminum powder:

$$Fe_2O_3 + 2\,Al \rightarrow 2\,Fe + Al_2O_3$$

As you can see, the iron (III) oxide and elemental aluminum mixture, once ignited, produces elemental iron and aluminum (III) oxide. Seems like a simple swap, right? Well what's not accounted for in this equation is the initial introduction of heat to ignite the thermite and the intense amount of heat generated from the reaction itself; the fact that the reaction starts with solid metal powders and forms molten liquid metal products should help to drive the point home. It also might be unclear just where the electrons are transferring between reactants here, so let's take a closer look at the reaction and the oxidation states involved:

$$\overset{[+3][-2]}{Fe_2O_3} + 2\,\overset{[0]}{Al} \rightarrow 2\,\overset{[0]}{Fe} + \overset{[+3][-2]}{Al_2O_3}$$

The elemental aluminum reduces the iron (III) oxide to elemental iron by transferring its three available electrons; thus the elemental aluminum is oxidized to aluminum (III) oxide.

This reduction-oxidation (redox) reaction happens in a flash. The enthalpy (or change in energy) of the thermite reaction is about minus 850 kilojoules per mole meaning that it strongly favors the products (or, the right side of the equation) since their total energy is less than the total energy of the reactants. This energy difference is released into the surroundings in spectacular fashion. And yes, the potent composition known as thermite is little more than the interaction of rust and aluminum foil, with the introduction of some significant heat to get things going.

So why are these two common ingredients so useful for the thermite reaction? Other metal powder fuels like magnesium, zinc, or boron could be used, but aluminum has a lot going for it. Its high boiling point (allowing the reaction to reach very high temperatures) and low melting point (allowing the reaction to occur more quickly since it takes place in the liquid phase) are ideal for the reaction itself. Aluminum's low cost, ability to form a passivation layer (a barrier that's less susceptible to corrosion), and the relatively low density of the aluminum (III) oxide product that allows it to float on the molten iron (preventing contamination of a weld) make it ideal for commercial, industrial, and military use.

Other than the very common iron (III) oxide, other chemicals like copper (II) oxide, iron (II, III) oxide, and manganese (IV) oxide can also be used. Iron (III) oxide produces more heat, though iron (II, III) oxide is easier to ignite; introducing copper or manganese oxides also eases the ignition process.

Now that we know what goes into making thermite, how likely would it have been for Walt to make his DIY thermite lockpick out of a Sketcher-roo? Let me dissuade you of the notion that these kids' toys are packed full of thermite; they're not ... that would be insane. The inside surface of the glass line-art toy is, however, coated in aluminum powder. Drawings appear as you turn the toy's knobs and the stylus inside the device scrapes a path through the powder. The lines appear black simply because there is no light inside the toy itself. Then, when you turn it upside down and shake it, polystyrene beads in the device—which provide that nostalgic shaky sound, as I mentioned earlier—redistribute the powder back onto the glass, "erasing" the earlier drawing. (These beads would have to be separated out from the aluminum powder, so I'll just assume Walt did that off screen.) Walt could have simply used bits of aluminum foil to prepare his thermite, but it's a much cleverer narrative turn to deconstruct the beloved Sketcher-roo.

If you need more reasons not to try this at home, the aluminum powder would likely be an inhalation hazard due to how fine it is, necessitating facemasks at the very least. Plus, you'd probably need far more than the ten or so Sketcher-roos Walt gathered in ordered to get enough aluminum powder for the thermite, up to two hundred toys, according to some sources.[2]

As for Walt and Jesse's attempt, storing their thermite powder in a plastic baggie wrapped in duct tape would probably have resulted in a fiery mess and an intact lock at the end of the day since thermite cutting and welding uses a ceramic container to keep the powder in place during the reaction. Also, Walt and Jesse openly stare at the white-hot thermite (without goggles) and then touch the handle to open the door (with gloves, at least), even though the reaction reaches over 4000°F. Safety issues aside, like the fact that molten iron could have sputtered or even rocketed out of the reaction site, Walt's DIY thermite probably would have cut through the warehouse lock, easy peasy. Walt's story about the Gustav Gun, however, only partly measures up to history.

Nazi Germany's 80-centimeter-caliber railway gun Schwerer Gustav, or Heavy Gustaf, was the largest-caliber rifled weapon ever used in combat, the heaviest piece of mobile artillery ever built at 1,350 tonnes (metric tons). Heavy Gustaf fired the heaviest shells at seven tons with an effective firing range of around twenty-four miles. (Walt quotes it at 1,000 tons, with seven-ton shells, and an accuracy of within twenty-three miles, so, close enough.) Its first combat test came in June of 1942 during the Siege of Sevastopol. After one month, firing only forty-eight shells, Heavy Gustav had worn out its original barrel (totaling nearly three hundred shots, including testing), though the heavily fortified city and its naval battery had been destroyed.[3]

Whether the gun was dismantled or destroyed, and whether that was undertaken by Axis powers or Allied forces, remains to be seen as history is unclear on this point. I prefer to think that Walt's lone commando, equipped with only thermite and scientific knowledge, could have taken the thing out, but the gun's inefficient design likely proved its own undoing.

Side RxN #8: Walter White's Fraying Nerves

One of Walt's cruder uses of his scientific knowledge to save his skin actually ended up burning him quite badly. In Season 5, Episode 6: "Buyout," Mike uses a plastic zip tie to secure Walt to a radiator in order to keep him out of trouble. With few other options available to him, Walt chews through a coffee maker's power cord—after turning off the power strip it's plugged into—strips the wires, threads them through the zip tie, and flips the power back on in order to burn through the plastic restraint with the resulting electrical arc. He earns his freedom after suffering a pretty gnarly burn on his wrist, but would this MacGyver move have worked in the real world?

Plenty of internet sleuths have offered up alternatives to Walt's shocking means of escape, but this discussion concerns what he actually did, not what he could have done. (For example, science genius Walt could have burned through the part of the zip tie attached to the radiator instead of his wrist, but that wouldn't be nearly so dramatic.) Since Walt was basically creating a short circuit here, there's a near certainty that the circuit breaker in either the power strip or the building's fuse box would have tripped as soon as the wires came into contact with each other. However, if the electricity was allowed to arc between the wires' ends, it certainly could have cut through even the police-issue zip tie; you can see this in action on the show itself because Walt's escape sequence was actually done using practical effects.

Inside *Breaking Bad*: On the Hollywood side of things, Bryan Cranston was indeed zip-tied to the radiator for much of the scene, but for the breakout itself, they used a lifelike prosthetic arm as a stand-in. The show's special effects foreman Ken Tarallo explains that a 12,000V transformer, which they controlled off camera, was used to provide the electrical arc, not a 120V household outlet.[4] (Narratively, Walt is obviously familiar with the concept of a "spark gap," having used it earlier in the series to start the fire that burned down the Superlab. Gilligan confirmed as much on the related podcast episode.) The zip-tie cuff was real, the same as the ones used in law enforcement, and it was one of the strongest they had available. However, the production team only used half of it to tie one of Walt's hands to the radiator; they're normally two-handed cuffs, of course.

X Corrosives:

Hydrofluoric Acid

From Dr. Donna J. Nelson:

In addition to having a love and fascination for science, Vince Gilligan is playful. These characteristics manifest themselves in *Breaking Bad*. There are also some off-camera examples of this.

One was his response when I sent him a t-shirt from the Student Affiliates of our local section of the American Chemical Society (ACS)—the Oklahoma Section (see figure 10.1). The students had created and sold t-shirts that bore the slogan "Chemistry—We do stuff in lab that would be a felony in your garage. Student Affiliates of the American Chemical Society" on the back. On the front, it merely stated "SAACS Oklahoma 2007–2008." I thought he might enjoy having one of these, as well as increase his awareness of the ACS. He obviously enjoyed it and played with it; he had photos taken of him wearing it in his Burbank office and emailed them to me.

Another obvious example is seen in his facial hair. When I first met Vince in 2008, he had no facial hair, but later on, by 2011, he had started growing it. At that time, I commented about his facial hair to some of the women crew, and they said "Look at the other men on the crew." I was astonished; most of them had this same style of facial hair, even the ones that handled props and managed the set. I asked, "Why are they doing that?" The response was "We don't understand it. We think it is a guy thing." That is right, because most women don't grow it. But also, most men on the set were emulating this characteristic of Walt!

I'm sorry, what were you asking me? Oh, yes, that stupid plastic container I asked you to buy. You see, hydrofluoric acid won't eat through plastic; it will however dissolve metal, rock, glass, ceramic. So there's that.—Walter White, Season 1, Episode 2: "Cat's in the Bag"

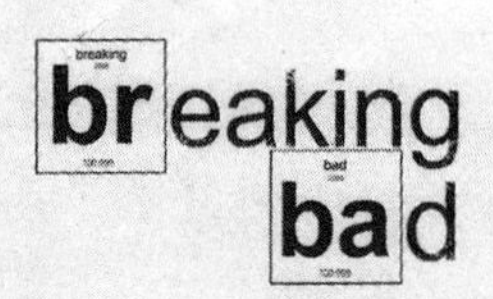

August 26, 2008

2501 W. Burbank Blvd., Suite 206
Burbank, CA 91505
(818) 841-0695

Prof. Donna Nelson
The University of Oklahoma
Department of Chemistry and Biochemistry
620 Parrington Oval, Room 208
Norman, Oklahoma 73019-3051
(405) 325-4811

Dear Donna,

Thank you for the fantastic t-shirt! I'm wearing it with pride. We so appreciate all your assistance this season. From pointing out things a non-chemist would never think of (like the likely purity of our barrel of methylamine) to working out the volume of Walt's end product, you have truly been an invaluable help to us. You will certainly get a few more questions from us before our season wraps up!

Best regards,

Vince Gilligan

Breaking Bad - Bldg. A
Albuquerque Studios
5650 University Blvd. SE
Albuquerque NM 87106
Phone 505.227.2700
Fax 505.227.2740

Figure 10.1
Letter from Vince Gilligan to Dr. Donna J. Nelson. Image courtesy of Dr. Donna J. Nelson.

So far, I've covered explosive problem-solving and a highly exothermic method of breaking and entering; now, the chemistry talk continues with one of the most recurring chemicals appearing throughout the entire run of *Breaking Bad*: hydrofluoric acid. Even the name of hydrofluoric acid (HF) conjures up images of mad scientists in their laboratories surrounded by sparking machinery, smoking concoctions, and other Hollywood special effects. The effects on display in *Breaking Bad*, as they pertain to HF, are decidedly more gruesome than that since Walt and Jesse often use the acid to dispose of dead bodies. It was used in Season 1, Episode 2: "Cat's in the Bag" on Emilio Koyama's corpse (and presumably his cousin Krazy-8, too); Season 4, Episode 1: "Box Cutter" to dispose of Gus's henchman Victor in the Superlab; and Season 5's Episode 6: "Buyout" and Episode 8: "Gliding Over All," following the shocking deaths of innocent bystander Drew Sharp (and his dirtbike) and fan-favorite fixer Mike Ehrmantraut, respectively.

Let's continue our conversation on chemistry in order to build on the earlier introduction and figure out just what makes acids so "mad science-y." Coming from *acidus*, the Latin word for "sour," acids provide a distinct flavor in foods like lemons (citric acid), vinegar (acetic acid), and even sour beers (lactic acid), but in works of fiction they're almost always used to dissolve everything from bank vault locks to bodies. Is there a precedent for that behavior in the real world? And if so, what gives acids the ability to do this?

By definition, an acid is a molecule or ion that is capable of donating a proton or hydrogen ion (H^+). Acids that more readily give up their hydrogen ions, or completely dissociate, are known as "strong" acids, while those that only partially dissociate are known as "weak" acids. For example, hydrochloric acid (HCl) will completely (100 percent) dissociate into H^+ and Cl^- ions in water since it's a strong acid; acetic acid (CH_3COOH) dissociates into H^+ and acetate, its conjugate base (CH_3COO^-), but only about 0.4 percent of a solution of this weak acid will dissociate. It's the hydrogen ion reacting with things like metals, cellulose in paper, or proteins in muscle and skin that breaks the materials down, so if there are more hydrogen ions readily available in solution, as is the case in strong acids versus weak ones, decomposition occurs more thoroughly and more quickly.

HF became a reliable way to eliminate the evidence of Walt and Jesse's increasingly dangerous and deadly dealings, but does the potency of the show's favorite acid add up in the real world? Oddly enough, HF might be more damaging to a body that's still alive than one that has shuffled off this mortal coil. Contact with the acid has the potential to cause deep, initially painless burns that result in tissue death. HF also interferes with the body's calcium metabolism, causing eventual system-wide toxicity, cardiac arrest, and even death. The gaseous form of the acid can immediately and permanently damage the lungs and corneas of the eyes. It's used in its diluted form in household cleaning products like rust stain removers and wheel cleaners, and, in its more concentrated form, for the manufacturing of refrigerants and in glass and metal etching.

It's nasty stuff, all right, but is it potent enough to eat through not just a dead body, but a gun, a ceramic tub, and the floor beneath it, as well? Given enough time, a high enough concentration, and sufficient volume ... maybe, but not in the way *Breaking Bad* portrayed it. It seems Walt and Jesse would have had to find another way to cover up their killing sprees. To understand why, I'll be breaking down HF's ability to break down organic material.

Advanced

Perhaps surprisingly, hydrofluoric acid is actually a weak acid since it only partially dissociates in water. That's understandably confusing considering that other halogens—the elemental group to which the fluoride component of hydrofluoric acid belongs—form strong acids: hydroiodic acid (HI), hydrobromic acid (HBr), and hydrochloric acid (HCl). So why is HF a departure from this trend?

Acid strength is determined by a few factors: electronegativity, atomic radius, charge, and equilibrium. "Electronegativity" is a term for how strongly an atom pulls electrons toward it; the more electronegative an acid's conjugate base is, generally, the more acidic the acid. Electronegativity tends to increase as you move from left to right across the periodic table, but it also decreases as you move from top to bottom in a column. This is due, in part, to an atom's size, which increases when moving top to bottom, making it more difficult for the atom to pull electrons closer into it.

Acid strength, however, also increases with an increase in the atomic radius, or the size of an atom's electron cloud. Generally, the larger the radius, the stronger the acid. This is because the charge of an atom, molecule, or ion is dispersed over the entirety of the electron cloud; a relatively larger electron cloud spreads that charge out more, reducing the charge density. In the case of negatively charged conjugate bases, increases in the size of the electron cloud lessen the bases' attraction to hydrogen, allowing the acid to dissociate more easily. Another way of saying this is that larger atoms, molecules, and ions are better able to give up H^+ due to a more stable distribution of the resulting negative charge. A larger atomic radius also makes for longer bond lengths between atoms, which are generally easier to break than shorter bonds.

To sum up those two points more simply: Acidity increases across the periodic table from left to right with increasing electronegativity, and down the periodic table with size. Additionally, more positively charged species also tend to be more acidic. Atoms or molecules with a positive charge, like hydronium (H_3O^+), can readily give up a proton and maintain stability. This is why hydronium is more acidic than water, which is more acidic than a hydroxyl group (OH^-). The accumulation of negative charges makes it increasingly difficult to give up more H^+.

Finally, the strength of an acid can also be represented by an equilibrium constant known as the acid dissociation constant (K_a) and, commonly, by its logarithmic counterpart, pK_a. If K_a is greater than 1, you're dealing with a strong acid; the larger the number, the stronger the acid. And the more negative pK_a is, the more readily the acid dissociates at a given pH, so the stronger it is.

That's a lot of terminology to take in all at once, and HF actually bucks a number of these general trends, so let's break the acid down by each factor:

Fluorine is actually the most electronegative element on the periodic table; the value increases as you move from left to right and bottom to top across the table. You'd be forgiven for thinking that HF should be one of the strongest acids out there, but the other factors complicate things a bit.

The atomic radius of fluorine is by far the smallest of the halogens and one of the smallest of the elements. Its negative charge is spread over a relatively small region of space when compared to the other halogens. This results in a short, strong bond between the hydrogen and fluorine atoms, which makes it harder for the H^+ to dissociate in water, therefore making it a weaker acid.

Additionally, HF's conjugate base (F^-) is very unhappy and unstable with a single negative charge spread across its small ionic radius. In other words, HF is more stable than the molecule's dissociated state.

Size and stability trump electronegativity when it comes to acid strength. Because of this, HF's equilibrium tends to shift toward the left, keeping the hydrogen and fluorine atoms together; its pK_a value is 3.2, landing it within the –2 to 12 range for weak acids. So while fluorine may be the most reactive of the halogens, hydrofluoric acid is relatively weak compared to the other hydrogen halides.[1]

Now don't hear the term "weak acid" and think that HF isn't dangerous; it certainly can be. Even a partially dissociated weak acid can dissolve materials like the ones seen in *Breaking Bad,* especially as more hydrogen ions are available in increasingly concentrated solutions. But hydrofluoric acid was an odd choice for the writers to use in the show as a method of body and evidence disposal because of its limitations. There are plenty of other options for powerful acids out there that they could have used instead, and to fully understand how nasty they can be, we need to talk a bit about pH and the Hammett acidity function.

The most common shorthand for representing an acid's potency is the pH scale, a negative logarithmic measurement of the "potential of hydrogen," which is a fancy way of accounting for the amount of hydrogen ions in an aqueous solution. (Free H^+ doesn't exist in water since the ions bond with the lone pair of unshared electrons on the water molecule's oxygen atom, so it's really hydronium ions we're looking for.)

The range of pH values runs from 0 to 14, with pure water being neutral and having a value of 7. Acids are found between 0 and 7 while bases run from 7 to 14. Values are derived from the compound's molarity (M), or a measure of how concentrated the solution is. The more acidic a solution is, the more hydrogen ions it's able to release into solution, and therefore the lower its pH value. And since it's logarithmic, a pH value of 1 is 10 times more acidic than a pH value of 2, and 100 times more acidic than a pH value of 3. Strong acids include sulfuric acid (H_2SO_4), which is found in drain cleaners and is also known as battery acid; hydrochloric acid (HCl), that is, stomach acid; and nitric acid (HNO_3), or "aqua fortis."

When it comes to truly scary acids, even the famous pH scale can't hope to measure their dissociative power. The strength of so-called "superacids," which are even stronger than 100 percent pure sulfuric acid, are measured by the Hammett acidity function (H_0) to account for very concentrated solutions. On this scale, pure sulfuric acid would have a value of –12, while superacids like the carboranes [$H(CHB_{11}Cl_{11})$] and fluorosulfuric acid ($HFSO_3$)] each have a Hammett acidity function of –18; the strongest acid currently known to man, fluoroantimonic acid, has been measured between –28 and –31.3. That makes fluoroantimonic acid roughly $2x10^{19}$ or twenty quintillion times more potent than the already dangerous sulfuric acid. This nasty piece of work will dissolve glass and plastic, and it even reacts violently in water, making it necessary to be diluted in a solution of hydrofluoric acid to keep it stable.[2]

With superacids like this truly approaching "mad scientist" levels of dissolving strength, you might be wondering how on Earth such a solution could be stored. Luckily, chemistry has the answer! The single bond between carbon and fluorine is nearly the strongest in organic chemistry—the single bond between silicon and fluorine is the only one that's stronger—and when more fluorine atoms are bonded to the central carbon, the bonds become shorter and stronger. Combinations of carbon-fluorine bonds create some of the least reactive organic compounds known to man, like polytetrafluoroethylene (PTFE), better known by its brand name, Teflon. That's right, the chemical that keeps food from sticking to your pans during cooking is also the only material that can safely store the world's strongest superacid.

Synthetic materials like Teflon and plastics are polymers, or chains or networks of repeating monomer units all linked together. This structure not only provides strong bonds between monomers, it also lowers the compounds' overall bond energy and makes them nonreactive. So Walt's insistence that Jesse uses a plastic tub to contain the evidence (and acid) checks out when it comes to real-world science; it may not have been good enough to hold certain superacids, but it's more than good enough for HF.[3]

All this talk of super-strong superacids might make the hydrofluoric acid seen in *Breaking Bad* feel a little less impressive, but that doesn't make HF any less dangerous to be around. Since it's so dangerous to handle and is used mostly in industrial purposes, I doubt that Walt's high school had a ready supply of it in the quantities needed to dissolve the bodies of Emilio and Krazy-8. But, for the sake of argument, let's assume it did. Would it be potent enough to (mostly) do away with any evidence?

Luckily the *MythBusters* team put this one to the test as well.[4] They revisited perhaps the most memorable use of hydrofluoric acid on the show, the scene in which the acid partially decomposes Emilio's body and dissolves his gun but also eats through the ceramic tub and the floor beneath it. This results in a grisly bit of clean-up work for Walt and Jesse when the body crashes through to the floor below. The *MythBusters* team ran a small-scale test using ceramic-glazed cast iron as a stand-in for a full tub, alongside drywall/plaster and wood for the floor, metal for the gun, and pig bone and flesh for the body. Under controlled conditions at the University of California, Berkeley chemistry lab, the samples soaked in 100ml of hydrofluoric acid for eight hours. Though the drywall, ceramic glaze, and pig flesh were partially dissolved, none of the samples were completely decomposed, busting the *Breaking Bad* science.

Their full-scale ramp-up opted for a stronger acid: "almost pure" 96 percent sulfuric acid with an additional ingredient that they kept secret. They probably used what's colloquially known as "piranha solution": a mixture of sulfuric acid (H_2SO_4) and 30 percent hydrogen peroxide (H_2O_2). This is a strong oxidizing agent used to clean organic residues from surfaces, to etch circuit boards and even to clean glassware in the lab (though usually this is reserved for porous sintered or "fritted" glassware). After taking the proper precautions—like carrying out their experiment in the middle of nowhere and hiring a properly trained hazardous materials company to do the dirty work, **so don't even think about trying this at home**—the full-scale test resulted in a lot of smoke (water vapor, carbon dioxide, and hydrogen gas) and a mostly dissolved pig (except for its bones), but an intact tub and floor, double-busting the show's fictional display of science. A triple bust came when they swapped fiberglass in for ceramic-glazed cast iron for their bathtub material and covered a pig in thirty-six gallons of their special acid combo. The pig completely dissolved in only a few minutes, but the bathtub and floor remained intact.

Perhaps if the tub had been made of a mild cheese, per Vince Gilligan's suggestion in the *MythBusters* special, the whole thing would have measured up a bit better. But if the writers really wanted to show art imitating life in terms of the "body disappearing" business, they should have opted for a strong base like lye to get rid of the evidence. The process of saponification, or soapmaking, has long used such strong bases to break down oils and fats similar to those comprising the human body. It's also what Mexican-American drug cartels use.[5]

With the right chemicals, equipment, and know-how, a body can be reduced to an oily, tan liquid in just about three hours. This process, known as alkaline hydrolysis, is also performed as an entirely legal way (in some states) to dispose of human remains as an alternative cremation process. So when it comes to disposing bodies, acids might do the job more completely, and they might be more recognizable to general audiences, but bases get it done quickly and with style.

Biology

Chemistry and physics are two of the "hard sciences" that work quite well in a lab setting because they're based on principles that can be demonstrated in practice; they also happen to make for explosive television. They are the cornerstones of *Breaking Bad*'s success since those demonstrable principles allow Heisenberg to make meth in the first place and also give him the tools to life-hack his way out of trouble. But it's when the hard sciences interact with the living world that things tend to get messy. You can take this in the literal sense, best exemplified by Emilio's run-in with the chemistry of phosphine gas and hydrofluoric acid, or Gus Fring's fatal physics lesson when he came face to face with an ANFO bomb. However, *Breaking Bad* succeeds equally well at delivering real-world science when exploring the complicated study of biology that is part and parcel to life as we know it here on Earth.

Walt's cooking sessions and explosive solutions may be the most memorable, but it's *Breaking Bad*'s focus on the "softer sciences" that helps to develop the show's flesh-and-blood characters; that's what I'll be covering in this part. For example, Walt's battle with cancer—a catch-all term for a number of diseases that have been under nonstop scientific scrutiny for decades—is every bit as prevalent in the story as Heisenberg's meth making. What's interesting is that both subplots deal with drugs and their effects on the human body, whether they be medicinal or recreational. And Walt isn't the only one struggling with health issues; Walt Jr., diagnosed with cerebral palsy at an early age, faces a lifelong struggle with a disorder that can't be cured simply by throwing money at it. (Actor RJ Mitte does a wonderful, earnest job of portraying Walt Jr.'s daily struggles since he has CP himself, though it's a less severe occurrence of the disorder.)

When it comes to problem-solving, Walt knows that not everything can be solved by explosions and corrosion; some things need a lighter touch. That's where his knowledge of chemistry and biology come to bear since the interface of those two disciplines is at the heart of toxicology, or in layman's terms, the study of poisons. (You might be surprised to find out just how many characters on this show tried to poison each other in increasingly creative ways.)

Then there's the fact that every single one of Walt's decisions actually affects the people around him, resulting in both physical *and* psychological damage. *Breaking Bad* does a solid job of dealing with the fallout of Walt's machinations in an honest way by portraying symptoms of post-traumatic stress disorder and introducing the notion of a fugue state to the narrative. Including all of these aspects isn't just good storytelling, it's also good science. I'll address each of them in the following chapters to see just how close *Breaking Bad* biology hews to the real-world science.

XI Psychiatry:
Fugue States, Panic Attacks, and PTSD

From Dr. Donna J. Nelson:

There are obvious differences between the Hollywood community and the scientific community, as seen in the personalities and activities of the people there. Most of us are familiar with Oscar Night, the red carpet, and the tabloids. But here I discuss the two communities' similarities and differences in the pursuit of their crafts.

The most noticeable similarity is in their goals; both communities seek to achieve excellence and perfection. However, the methods used to achieve that are so different that throughout my interactions with *Breaking Bad* actors and crew, it seemed I was in a different world.

First, the vocabulary was different. They wanted to learn mine in order to achieve Vince's goal of science accuracy. I had to learn theirs in order to help them.

It was obvious that they were incorporating the science they learned from me into the show. I was sensitive to the fact that they didn't know typical science jargon, so I gave them information that pertained to their show and avoided overwhelming or confusing them with jargon. However, occasionally, scientific terms that were not immediately pertinent slipped through, such as when I used the word "precursor" and Vince adopted it. Other times they weren't interested, as I mentioned earlier with the word "stoichiometry."

The way in which we influence the outcome of work was different. Scientists exclude their opinions as they gather data; directors get dozens of "takes" of a scene.

Our attitudes about selecting the outcome of the work were opposed. They would look at all the "takes" of a scene and select the one they liked the best. We would call that cherry-picking and biased; scientists would forbid it.

Our goals and considerations were different, as when I described their selecting a reagent based on how easy it was for the actors to say.

However, I always aligned my goals with theirs, in order to support their creative writing. I was flexible and adaptable, so I learned a lot and had fun!!

I think some of these differences were responsible for a myth I was told existed in Hollywood. There was a myth that it was impossible to have a hit show *and* a science advisor. Supposedly the science advisor would constantly tell the crew, "You can't do this; you can't do that"—and so on. The effect was to try to make the show a science documentary and remove the artistic license and the creativity. The show would then not interest the audience and would fail. When I heard that, I knew my main goal had to be to align my goals with those of the crew.

Because of that, when editing script pages, I took a minimalistic approach—change as few words as possible and still get the science right. I needed to respect what the writers had written. After all, they knew how to write for viewers. It was more than just respecting alliteration and cadence; if a three-syllable word starting with p was wrong, I tried to replace it with a three-syllable word starting with p that corrected the science. This can be harder than writing a totally new text, but this was what I did, and I got on fabulously with the crew.

There was no fugue state. I remember everything.—Walter White, Season 2, Episode 3: "Bit by a Dead Bee"

It's not all cold-hearted chemistry and explosive physics at work in *Breaking Bad*. The reason that viewers get so invested with Walter White and his slow descent into villainy has much more to do with character interactions than chemical reactions. Walt's increasingly violent and selfish choices come with unavoidable blowback that affects not only Walt himself but also the people around him, friends and enemies alike. Sometimes these effects manifest physically, as is often the case for antagonists who cross Walt's path and are permanently put to rest, but those who survive his villainy are left with their own invisible scars from emotional trauma.

There's a psychological tension that runs throughout the entire series, and it's portrayed in different ways. Sometimes, as in his infamous fugue state, Walt uses his odd and questionable behavior as a means to an end in order to bail him out of trouble or provide an alibi for his criminal misdeeds. And other times, innocent bystanders are caught in the crossfire, often with devastating psychological effects. *Breaking Bad*'s most extreme example of the far-reaching consequences of PTSD (post-traumatic stress disorder) claimed the lives of 167 people, thanks to a domino effect that began with indecision on the part of Walter White, which led to the death of Jane Margolis (the tattoo artist who was Jesse's landlady and girlfriend), and ultimately ended with her father Donald Margolis, a depressed and distracted air traffic controller, making a costly mistake.

While we never spend any screen time with these 167 victims, we do spend a lot of time with two central characters who both suffer, directly and indirectly, at the hands of Walt. Hank's multiple panic attacks and Jesse's struggle with PTSD throughout the series are two additional ways the show brings attention to psychological trauma. Now let's explore these phenomena further.

Fugue States

101

While the psychological trauma that Jesse experiences is very real—more on that in a bit—Walt's own brush with a psychiatric disorder was 100 percent fabricated. In Season 2, Episode 3: "Bit by a Dead Bee," Walt needs a realistic excuse to explain his absence, having spent days in the desert under Tuco Salamanca's control, enough time out of contact to cause his family to put up "Missing" posters. Short of claiming he was kidnapped, got lost, or was moonlighting as a meth-making drug-dealer, Walt's next best option is to walk completely naked through a supermarket until the authorities pick him up and reunite him with his family at the local hospital. His ruse works just long enough to throw any suspicion out the window since his family's just happy to have him back, safe and sound. But do fugue states like the one Walt made up really occur?

Perhaps not surprisingly, they do! In what's known as a dissociative fugue, a type of amnesia, people lose some or all of their personal memories, usually caused by trauma or stress. These events can last from a few hours, to months, to possibly decades, and people suffering through it may travel far from home, assume a new identity, and even start a new life unawares.[1] Things get a little complicated when medical professionals attempt to separate a legitimate dissociative fugue, for which there is no direct physical or medical cause, from someone who's "faking it" in order to escape a frustrating marriage, dodge accountability, or avoid military combat, for example, but there are definitely ways to sort this out.

However, the uncertainties of the fugue state actually play right into Walt's hands. Dr. Soper and Dr. Delcavoli—Walt's attending doctor and oncologist, respectively—look into any potential side effects from his cancer medication or interactions between them, but find nothing to explain his behavior, so they refer him to psychiatrist, Dr. Chavez. Walt confesses to making up the fugue state, under doctor/patient confidentiality of course, as an excuse to simply get away from his family obligations for a few days. Clever guy, that Walt, and a great actor, too. He stops short of confessing his real crimes but manages to concoct a story that gives him a clean bill of health and covers his very suspicious behavior ... for a time. It's a good one-use plot device in the *Breaking Bad* narrative, but fugue states get much more complicated in the real world.

Advanced

Not a specific disorder itself, a dissociative fugue is considered to be under the dissociative amnesia disorder in the Diagnostic and Statistical Manual of Mental Disorders, or DSM-5. As hazy as the cause of real-world fugue states are, there are numerous documented accounts of them taking place: There's the 2006 case of a fifty-seven-year-old lawyer, husband, and father of two who left his Westchester County, New York garage one day and was found six months later in a Chicago homeless shelter living under a new name with no recollection of his former life; a combination of his experience as a Vietnam War veteran and narrowly avoiding the 9/11 attacks likely triggered a return of painful memories that caused his condition. In his book "The Medical Detectives, Volume II," Berton Roueché tells of a man who, in the mid-twentieth century, found himself in a New York hotel, unable to remember his name, after failing to show up at his father-in-law's store in Boston. Those details came back to him suddenly and all at once after many failed attempts to jog his memory.[2] Even mystery and thriller author Agatha Christie supposedly experienced a fugue state, having disappeared from her home in the winter of 1926 following the death of her mother and learning of her husband's ongoing affair. In fact, she had checked herself into a health spa under a different name (one she cheekily borrowed, in part, from her husband's mistress) and concocted a whole other story for why she was there, further muddying the waters regarding the cause of her disappearance.[3]

Symptoms of a dissociative fugue event include sudden travel away from home and habitual places without any ability to recall one's past, identity confusion, and either partial or complete assumption of a new identity.[4]

Causes of a dissociative fugue tend to be severe stress or an emotionally traumatic event, something so painful or disturbing that the mind opts to file that information away in a hidden folder somewhere, taking related memories along with it. That's an important distinction from other types of memory loss since a fugue state's forgotten information can be recovered. Doctors will usually be able to distinguish between a legitimate fugue and an illegitimate one through neuropsychological tests, since fakers tend to overdramatize and exaggerate their symptoms, and they often have more practical reasons for faking memory loss that better explain their behavior. (Let's just say it's a good thing Walter White inherited some of Bryan Cranston's acting skills.)

But much like Walt wasn't off the hook because he claimed to be fine after his fugue episode, real-world dissociative fugue events should result in a doctor's evaluation. Even when examining patients experiencing memory loss that seems to have no obvious physical cause, the first step is to look for a neurological cause, such as stroke, viral encephalitis, epilepsy, or head injury. Also, the loss of generic knowledge and memory points to a psychogenic or psychological origin instead of a physical cause. A functional MRI (magnetic resonance imaging) or PET (positron emission tomography) scan may also reveal underlying brain damage as a potential cause.[5]

The really difficult part of the fugue state isn't necessarily the fugue itself; the person may exhibit no symptoms beyond mild confusion. However, once the memories return and the realization sets in that not only were they in a fugue state but also they must now face the original cause of their stress, a patient can experience depression, grief, aggressive impulses, and even suicidal tendencies. Treatment for these feelings includes psychotherapy, hypnosis, and interviews facilitated with the use of drugs, like an intravenous sedative. This approach is more in line with helping the patient figure out how to respond better to the situations and emotions that trigger these events, and thus hopefully preventing more in the future. However, said treatments are less successful with recovering memories from the fugue period itself.

So Walt's relatively easy-breezy "recovery" period after his fugue state may have tipped off his medical professionals—and Skyler and Walt Jr., if we're being honest—that everything was not on the up and up. However, with panic attacks and PTSD, *Breaking Bad* wasn't faking it.

Panic Attacks and PTSD

101

Arguably, two of the most physically and psychologically damaged characters on *Breaking Bad*, as a direct result of Walt's actions, are Hank Schrader and Jesse Pinkman. Sure, numerous innocent victims and criminals alike have paid the ultimate price for crossing Heisenberg's path, and Walt's fictional family will likely carry the emotional trauma with them well after the scripts have ended, but Hank and Jesse bear the brunt of both physical and psychological violence throughout the series.

Hank, introduced as a macho, boisterous DEA agent as a foil for the mild-mannered Walter White, quickly rises through the ranks of law enforcement. From busting low-level meth-heads early on, to his assignment to the Tri-State Border Interdiction Task Force (before getting suspended temporarily), to his appointment as ASAC (Assistant Special Agent in Charge), to the point that he ultimately resumes his one-man mission to find Heisenberg, the career pressure on Hank rises exponentially as the seasons wear on. He also survives a shootout with Tuco Salamanca; a deadly explosion in Juárez, Mexico; another shootout with the Salamanca cousins; and the stunning realization that his brother-in-law is a meth kingpin. That's a lot to ask of anyone.

When picking bar fights, brewing beer, and studying minerals don't offer enough of a pressure release, Hank's inner anxiety manifests in debilitating panic attacks. Gilligan even references Hank's brushes with PTSD—specifically the panic attack in the elevator in Season 2, Episode 5: "Breakage"—as laying the groundwork for his car crash following his realization that Walt is in fact Heisenberg.[6]

Jesse, meanwhile, was perfectly content as the low-level meth-maker "Cap'n Cook" before Walt brought a world of hurt to his door. Over the seasons, Jesse has had to dispose of multiple bodies (including one kid ...); has witnessed first-hand the deaths of friends, loved ones, and enemies alike, and is guilty of taking lives himself; and has been beaten within an inch of his life by both drug dealers and law enforcement; not to mention his history dabbling in illicit drug use. All of these factors lead straight to PTSD. This trend continues throughout the rest of the series, even complicating the DEA's attempts to get Jesse to turn on Walt. (Jesse gets spooked in Season 5, Episode 12: "Rabid Dog," as a result of his PTSD when trying to lure Walt into a confession.) Show writer George Mastras believes that Jesse is clearly suffering from PTSD and said as much in the *Breaking Bad Insider Podcast* episode for Season 4, Episode 2: "Thirty-Eight Snub."[7]

Hank and Jesse have lived through some traumatic events, so to not have those events affect them in some profound ways would be a disservice to the show's basis in reality. Hank's panic attacks—sudden onsets of anxiety and fear that manifest physically through shaking, sweating, shortness of breath, and heart palpitations—are completely understandable considering the things he's seen and dealt with; Jesse's disturbing thoughts and visions, mental stress, and ratcheted-up fight-or-flight response are on par with people suffering from post-traumatic stress disorder following a particularly troubling experience. (Hopefully Jesse gets the help he needs wherever he ends up, but unfortunately Hank's beyond such therapy.)

To find out just how close to reality the psychiatric issues portrayed in *Breaking Bad* really were, our discussion will have to get a little more advanced.

Advanced

"Panic attack" is the colloquial term that has gained acceptance in recent years, but there's also the related "anxiety attack." Ricks Warren, PhD, a clinical associate professor of psychiatry at the University of Michigan at the time he spoke with the university's health blog reporter Kevin Joy about panic and anxiety attacks,[8] remarked that the attacks have "very different emotional conditions."

Anxiety can be understood as excessive worrying over impending events (death or illness), minor events (appointments), or general uncertainty. Normally a chronic condition, symptoms of anxiety include "fatigue, hyper-vigilance, restlessness, and irritability." It's what people experience while worrying about an event in the future and its potentially bad outcome; as such, anxiety attacks tend to come on gradually. Neurologically, these attacks are associated with the brain's prefrontal cortex, which handles planning and anticipation.

Panic attacks, meanwhile, are short, intense bursts of fear from a sense of immediate danger; the fight-or-flight alarm response is at work here. These symptoms, which can resemble a heart attack, include increased heart rate, chest pains, and shortness of breath, and typically last no longer than thirty minutes, though they can occur without reason. The body's autonomic nervous system (which includes bodily functions that are not consciously controlled, such as breathing, heartbeat, and digestion) and amygdala (an area of the brain involved in emotions, survival instincts, fear, and memory) are active here since their roles are to detect threats and respond to them accordingly.

People can experience both panic attacks and anxiety attacks simultaneously. Imagine walking down a dark alley (or don't, especially if you're already experiencing anxiety from reading this section); anticipating possible danger could lead to an anxiety attack. Now, God forbid, should something dark and dangerous actually confront you in that alley, a panic attack would be perfectly understandable. There's actually an effect dubbed "anticipatory anxiety," which accounts for people worrying about a future panic attack, resulting in a positive feedback cycle. People who suffer multiple panic attacks, as in chronic cases of anxiety disorders, may even begin avoiding the places where these attacks have occurred in order to minimize their anxiety and decrease the risk of future incidents.[9]

Throughout *Breaking Bad,* Hank never seems to show symptoms of having an anxiety disorder, but anxiety itself is often anonymous; most people won't even be able to spot when someone else is anxious. It's a normal, adaptive, and temporary aspect of our everyday lives and in no way a measure of "toughness."[10] According to the Anxiety and Depression Association of America, about six million American adults experience panic disorder annually, with women being twice as likely as men to do so.[11] The good news is that there's no need to suffer in silence or be embarrassed by it; the disorder is very responsive to treatment through counseling or even medication.

Hank's very specific panic attacks occur when his comfortable routine and worldview are shaken, if not shattered. His unexpected run-in with Tuco Salamanca, the subject of the DEA's manhunt at the time, ends with the drug dealer's death at Hank's hands, a feat that gets him appointed to an elite task force. In a typical drama, this would be the law enforcement tough guy's dream scenario: Kill the bad guy and reap the rewards. In *Breaking Bad,* however, the real-world implications of taking a man's life in self-defense and the rushed promotion that puts extra pressure on Hank's shoulders is too much for him to adjust to in a healthy way. Then, when Hank, who's barely keeping it together at this point, enters into the confined space of the elevator, his fight-or-flight response goes into overdrive; he can neither flee from the tiny metal box, nor can he fight an enemy that's already been dispatched or has yet to manifest.

Things don't get any easier for Hank. In Season 2, Episode 7: "Negro y Azul," he finds himself a fish out of water in the El Paso office where no one appreciates his brand of humor and he literally doesn't speak the language of his cohorts. A minor panic attack here actually saves Hank's life since it takes him out of the deadly range of an improvised explosive hidden away within the severed head of an informant attached to a living tortoise. Hank's mental resolve takes some time to recover after this traumatic event and that of the shoot-out against the Salamanca cousins that almost claimed his life in Season 3, Episode 7: "One Minute." It's a testament to his commitment to his job that he soldiers on. The next major panic attack occurs in the Season 5 midseason finale, "Gliding Over All," when Hank fully realizes that Walter White, his mild-mannered brother-in-law, is actually Heisenberg. This "sky is falling" realization causes Hank to careen off the road while driving, owing to the overwhelming surplus of poorly managed psychological stress.

It's interesting to note that, when faced with actual danger, Hank is cool and calculating. It's only after surviving that danger that Hank's anxiety and panic kicks in. Dean Norris's performances throughout the show bring the very real symptoms of anxiety and panic attacks to life—rapid heartbeat, sharp chest pains, lightheadedness, difficulty breathing, difficulty focusing, trembling, sweating, and more—in terrifying fashion. His performance opposite Cranston also brings a whole new level to Walter White's intimidating line, "If you don't know who I am, then maybe your best course ... would be to tread lightly."

It's that simmering danger that Walt brings to *Breaking Bad* from the very beginning, and it's his partner Jesse Pinkman who often finds himself paying the price. Jesse's slow-developing PTSD builds on a series of increasingly traumatic events over time. He's able to rationalize some of these experiences, regarding them as nudges toward straightening his life out, but the first such event that pushes him over the edge has to be the death of Jane Margolis in Season 2, Episode 12: "Phoenix."

The resulting guilt trip sends Jesse into and out of rehabilitation facilities and on drug-fueled, downward spirals. The show's writers doubled down on that guilt by having Jesse kill Gale Boetticher in the Season 3 finale, "Full Measure," resulting in Jesse's attempts to escape those feelings by hosting endless, drug-fueled raves and house parties.

Unable to escape from Walt's machinations, Jesse becomes increasingly distrustful of his partner and more and more paranoid (rightfully so) by the day. His mental state also comes into question when he's found absent-mindedly tossing money out of his car into an Albuquerque neighborhood in Season 5, Episode 9: "Blood Money." After five seasons of witnessing Jesse's psychological decay from his proximity to death and violence, it's hard not to get emotional at the overall final shot of Jesse, laughing and crying exuberantly, once he is finally free.

It's no stretch to claim that Jesse suffers from PTSD. This mental disorder develops following exposure to traumatic events that range from threats to a person's life, to car crashes, near-death experiences, sexual assault, and the complicated arena of warfare. A host of disturbing thoughts, feelings, and dreams that are related to the event stem from it, along with mental and physical distress that can be triggered by related cues. Attempts to avoid these cues through changes in behavior and the way the person thinks and feels, and an uptick in the fight-or-flight response soon follow. That last one should sound familiar from our discussion of panic and anxiety attacks, which makes sense considering that PTSD can cause such attacks to occur. A difference here, however, is the longevity of such symptoms; PTSD effects can last for more than a month following the original event while similar effects occurring for less than a month could be diagnosed as acute stress disorder.

Officially recognized by the American Psychiatric Association in the third edition of the Diagnostic and Statistical Manual of Mental Disorders (DSM-III) in 1980, the term PTSD replaced the World Wars-era references to "shell shock" and "combat neurosis" following the Vietnam War. It was classified as an anxiety disorder in DSM-IV and reclassified again in the current DSM-V as a trauma and stressor-related disorder.

Following a physical examination and a psychological evaluation, a doctor may diagnose PTSD if the patient directly experienced the traumatic event or saw it occurring to others, or both, is repeatedly exposed to traumatic events' graphic details, and if symptoms last for more than a month, impair a patient's ability to function in society, and affect relationships in a negative way.[12] I'm no doctor, but it sounds like Jesse Pinkman's experiences fit that diagnosis.

Part of the long-term struggle with the disorder is an increased risk of self-harm and even suicide. Luckily, counseling and medication are available to help people suffering from PTSD. Psychotherapy, or "talk" therapy, can help improve symptoms by teaching the skills needed to address thoughts of self-harm and suicide in the first place and cope with them should they arise, encouraging a more positive outlook in general, and treating other issues such as depression, anxiety, and drug abuse. Patients can seek out one-on-one therapy, group therapy, or both. Different types of psychotherapy also exist, including:[13]

- Cognitive therapy, which helps to recognize patterns of thought that might keep a patient stuck in a feedback loop.
- Exposure therapy, a behavioral therapy that allows patients to safely face frightening situations and memories in order to cope with them.
- Eye movement desensitization and reprocessing, or EMDR, which combines exposure therapy with specific eye movements intended to help process traumatic memories and change how a patient responds to them.

Additionally, medication is available for patients suffering from PTSD:

- Antidepressants for help with symptoms of depression and anxiety while improving sleep and concentration. Examples include selective serotonin reuptake inhibitors (SSRIs) like the FDA-approved brands Zoloft and Paxil.
- Anti-anxiety medications, which alleviate severe anxiety symptoms.

The main understanding to take away here is that anxiety disorders and PTSD are very real problems in our society but are common enough that there are many options for treatment for anyone suffering from them. I like to think that Jesse, having escaped the deadly world of illegal meth making once and for all, went on to seek out the help he so desperately needed and is now making a living as a happy, healthy, wood-working artisan in Alaska.

Side RxN #9: Walt Tells the Truth

In the Season 2 finale, "ABQ," Walt goes under the knife as part of his cancer treatment, but he also goes under the effects of a chemical relaxant just before surgery. The fast-acting chemical, which is never named, soon has Walt feeling all right but also loosens his tongue a bit. Skyler asks where his cellphone is and Walt responds, "Which one?" That's a big narrative bomb of an admission of guilt considering that the Whites have had heated conversations about Walt's secret burner phone. But would a preoperative anesthetic really act as a truth serum?

If Walt had been undergoing a relatively short, noninvasive procedure, he could have been given a narcotic like morphine, Demerol, or Fentanyl. However, Walt is undergoing a surgery that opened up his chest cavity—as evidenced by the scar on his side that's shown in later episodes—so he goes under general anesthesia.[14] But before Walt is completely out of it, he goes through the procedure's premedication steps, which include the administration of midazolam, a fast-acting benzodiazepine that goes by the much easier-to-pronounce name of Versed. This sedative, which starts working within five minutes and can last as long as six hours, can cause short-term amnesia but not a sudden need to confess. Its side effects include low blood pressure, sleepiness, and a lessening in breathing efforts.

Such a sedative is not a truth serum, per se, which means law enforcement likely won't find any use for it in criminal interrogations.[15] However, presurgery sedatives do take away a person's control of their faculties, and some people respond to this disinhibiting effect by either "opening up" or simply talking a lot more than usual. That's probably the mildest form of disinhibited behavior; others include unwelcome sexual advances and outright hostility or violence. In Walt's own defense, he later claims that he could have said the world was flat while under medication, but even if he'd copped to being a meth maker, most anesthesiologists probably would have chalked it up to him watching too much TV.

XII Pediatrics: Cerebral Palsy

From Dr. Donna J. Nelson:

I was offered a cameo filmed during a set visit to the show on May 12, 2011. They asked, "Would you like to portray a nursing home attendant?" I was happy to do it. The cameo was to air on September 25, 2011, but it was cut; however, it can be viewed at the end of the Deleted Scenes file on Disk 2 of the *Breaking Bad* Season 4 Blu-ray (see figure 12.1).

It was in Episode 41 overall, Episode 8 in Season 4: "Hermanos." In this episode, Gus visits Hector in a nursing home and informs him of the death of his nephews after their attempt to kill Hank. My scene was the last scene filmed at that location, and it consisted simply of a camera panning across me as I wrote on a writing board in the entryway of the nursing home. A photo of me in my nursing home garb taken that day is shown below. By the way, notice the facial hair on the prop person in the background; another example of matching Walt's facial hair.

I recall putting on the outfit around 9 am that morning, but the actual filming wasn't done until late that afternoon. As I wandered around the set that day, I asked one of the makeup artists if they were going to put makeup on me, and the reply was "Honey, we do cuts and bruises and tattoos." Notice the bruises on Jesse's face in the deleted Superlab scene on Disk 2 of Season 4; the bruises certainly looked real, even in person.

An interesting characteristic that is common to all movie and TV sets I've seen is the food service provided. From my experience, it has almost everything one could want. However, if a person has a favorite food that is missing, they only need to request it, and it appears later that day or the next day. Often, the food is in food trucks, which one can explore. But at mealtimes, there is usually a fabulous buffet assembled, sometimes with a separate dining trailer provided.

In between shoots, the actors had to remain in character. Sometimes this was challenging for them. If a scene occurred in a room, the walls would each need to be removed, one by one, and one or more cameras moved in, so that takes from all angles could be obtained. Also, sometimes the ceiling would be removed, to get a shot from above. Replacing a wall and breaking out a different one could

Figure 12.1
Dr. Donna J. Nelson in her *Breaking Bad* cameo. Image courtesy of Chris Brammer.

> take a while. During this time, the actors had to remain in character. Sometimes Bryan, as Walt, walked around the set talking to us; to me he might say, "Have you read any good chemistry articles recently?" I might reply, "Yes, there were some good ones in the *Journal of Organic Chemistry*." And he would respond, "Good, we'll talk about them later."

101

My wife is seven months pregnant with a baby we didn't intend. My fifteen-year-old son has cerebral palsy. I am an extremely overqualified high school chemistry teacher. When I can work, I make $43,700 per year. I have watched all of my colleagues and friends surpass me in every way imaginable. And within eighteen months, I will be dead. And you ask why I ran?—Walter White, Season 2, Episode 3: "Bit by a Dead Bee"

Walt's recurring battles with lung cancer throughout the entirety of *Breaking Bad* get the lion's share of the focus when it comes to medical maladies.

However, it's RJ Mitte's character Walter White, Jr. who deals with the very real struggles of cerebral palsy on a daily basis without a fraction of the sympathy his father gets. That's in part a testament to the central story of *Breaking Bad*—it is a tale of Walt's downfall, after all—and in part due to the strength of character and conviction of Mitte himself.

Many *Breaking Bad* fans are aware that actor RJ Mitte has cerebral palsy himself. However, fewer fans may know that Mitte's CP impairs his movement less than Walt Jr.'s affliction does. For example, Walt Jr. relies on crutches to help him move and has a more pronounced speech impediment, but Mitte actually had to train himself to play up both of these aspects in his performance.[1] It's refreshing to see an actor who has dealt with his condition just about every day of his life playing a role that treats CP honestly, a role that just as easily could have gone to an able-bodied actor who would literally be acting out symptoms they've never felt.

Those symptoms include a variety of coordination and movement impairments that can affect an individual's communication and mobility. A few plot points specifically revolve around Walt Jr.'s CP—When he is picked on by bullies at the clothing store, when his father Walt tells his psychiatrist that his son's cerebral palsy is a burden, and when Walt Jr. is figuring out how he's going to pass his driver's test—but perhaps it's preferable to simply portray Walt Jr. as any other American teenager: Bouts of rebellion against his parents, getting busted for underage drinking and presumed marijuana use, and getting his first job at the family business (the car wash his parents purchased for money laundering, not the meth lab).

RJ Mitte and the *Breaking Bad* team do a fantastic job of portraying cerebral palsy earnestly without oversympathizing with the disorder. But to find out more about what makes CP such a difficult thing to deal with, our discussion's going to have to get a little more advanced.

Advanced

What Is Cerebral Palsy?

Though it wasn't known by the name "cerebral palsy" until Canadian-born physician William Osler coined the term in 1888, this affliction has affected mankind for thousands of years.[2] The blanket term "cerebral palsy," referring to the brain's cerebrum (which regulates motor function) and the paralysis of voluntary muscle movement, covers a group of movement and posture disorders appearing in early childhood.[3]

It's the most common childhood physical disability, with 1 in 500 babies being diagnosed with CP and 17 million currently afflicted worldwide. There are four main types:[4]

- Spastic—Caused by damage to the brain's motor cortex, resulting in stiff, exaggerated movements, and accounting for more than 70 percent of cases.
- Athetoid/Dyskinetic—Caused by injury to the balance and coordination center, the basal ganglia, resulting in involuntary tremors.
- Ataxic—Damage to the cerebellum (where the brain meets the spine), resulting in a lack of coordination and balance.
- Mixed—Multiple types of CP occurring within the same individual.

What Causes It?

Cerebral palsy research and treatment fall within the fields of pediatrics and neurology since CP occurs from damage to the fetal or infant brain, either before, during, or within five years of birth. This damage prevents normal development of the brain, especially in the areas that control motor function, and shouldn't be confused with damage to actual muscles or nerves themselves. As for how the damage occurs, common causes include bacterial or viral infections; bleeding or hemorrhaging in the brain; a lack of oxygen to the brain, or asphyxia; prenatal exposure to drugs, alcohol, mercury poisoning, or toxoplasmosis; birth injuries; and head injuries. Infants born prematurely face a higher risk of developing CP. Other risk factors include low birth weight, breech births, maternal diabetes, high blood pressure, and poor health. Twenty percent to 50 percent of cases are estimated to have unknown causes, however, so clinical trials continue research in this area.[5]

What Are the Symptoms?

Symptoms vary among individuals but include impairment to body movement; muscle control, coordination, and tone; reflexes; and balance. People with cerebral palsy may also have sensory and intellectual impairments, difficulty speaking and learning, and epilepsy. Though people with CP may require mobility aids or speech-and-hearing-assistive devices, or both, the majority of them can still walk and verbally communicate.[6]

Parents should pay close attention to whether their child is meeting or missing important developmental milestones since they may serve as early warnings for cerebral palsy. Common warning signs include stiff muscles, jerky reflexes, lack of coordination, and drooling, among others. Since CP is an umbrella term for a group of disorders, other brain developmental conditions may coexist with it, including Autism Spectrum Disorder and Attention Deficit Hyperactivity Disorder, among the other impairments and disabilities already mentioned.

What Is the Treatment for CP?

Following notable delays in a child's development between eighteen months and five years from birth, doctors can use imaging tests like magnetic resonance imaging (MRI), computed tomography (CT) scan, electroencephalogram (EEG), and cranial ultrasounds to further diagnose cerebral palsy. Unfortunately, there is no cure-all for CP, but with early-in-life treatment and lifelong management, the disorder's impairment of the individual's life can be minimized.

Since a child's brain and body heal, adapt, and recover more quickly than do those of older patients, a focus on nurturing development at an early age is stressed. This helps a child with CP learn to be self-sufficient and lead a normal, satisfying life when they eventually mature into adulthood. A team of specialists trained in traditional therapy, medications, and surgical procedures can work together to help a patient with CP improve their motor skills.

How Severe Is RJ Mitte's CP?

Mitte, who was adopted at only a few weeks old, showed no signs of CP until he was about three years old. Tightness in his body's tendons forced him to walk tiptoe and his fingers grew rigid. Eventually Mitte and his adoptive, single mother learned that he was born through an emergency caesarean and had to be resuscitated at birth since he wasn't breathing; Mitte suffered permanent brain damage from the experience.

After his diagnosis, Mitte had to go through a painful six months of having his feet repeatedly bent and put into casts in order to straighten them out. Other treatments in his childhood included speech therapy and exercises designed to improve hand-eye coordination. It was actually Mitte's time spent playing sports in school that allowed him to avoid surgery and set aside his leg braces; a few short years later, he found himself practicing using his crutches again with friends at his local Shriners Hospital in order to better play the part of Walter White, Jr. on *Breaking Bad*.[7]

In terms of his continuing treatment—since CP is a chronic, life-long disorder—Mitte fights off leg stiffness with yoga and uses a prescription muscle relaxer. Perhaps of equal interest and importance is Mitte's work as an ambassador for CP awareness through United Cerebral Palsy.[8] He also takes on speaking engagements to educate people on how to overcome adversity and turn disadvantages to advantages by sharing his personal experience of his transformation from the target of schoolyard bullies into a celebrity TV star.[9] And for as much as Mitte's performance on *Breaking Bad* did to bring CP into the spotlight and show the range that actors with disabilities could bring to the table, Mitte knows that there's still a lot more work to do. Here's hoping this section of the book helps that effort in some small way.

Side RxN #10: Walter White's Heredity

Cancer gets a lot of the focus in *Breaking Bad* since Walt's diagnosis incites him to begin making meth. Secondary to that but still very present throughout the show is Walt Jr.'s daily struggles with cerebral palsy. However, a third debilitating disease pops up in Season 4, Episode 10: "Salud." In this hour, Walt is telling Walt Jr. about his own father, who passed away when Walt was only six years old due to complications from Huntington's disease.

Walt has many narrative reasons for telling his son this story, be it manipulation, a heartfelt father/son moment, or maybe even an attempt to justify his own shortcomings, but the disease in question is interesting from a biological perspective as well. Huntington's disease, which has been recognized for hundreds of years but not sufficiently described until George Huntington's 1872 study of its pattern of inheritance, results in the death of brain cells, which causes mood instability, impairment to mental abilities, a lack of coordination, and jerky body movements, all of which worsen over time. Most patients inherit the disease, though rare cases exist of a person having the disease without either parent being afflicted.

Since Walt's father had the disease—and we're assuming his mother did not—Walt had, at best, a 50 percent chance of inheriting the afflicted gene. (The autosomal dominant pattern of inheritance means that only one copy of the mutated gene on a non-sex chromosome is needed to inherit the disease.) The mutation produces abnormal proteins that gradually damage brain cells, though the mechanism isn't fully understood and no current treatment can halt or reverse the course of the disease.[10]

As for testing, Walt says they ran tests on him when he was a kid, but he "came up clean." This would have been in the 1960s, most likely, well before modern genetic testing. Also, since symptoms tend not to develop until later in life than childhood, the patient may be without symptoms, or presymptomatic, for years. However, research in the 1950s and the start of the Hereditary Disease Foundation in 1968 did advance the field and pave the way for future testing and understanding. Predictive genetic testing for Huntington's via a blood test only became available after 1993.[11]

Walt also says that his father became very ill when Walt was four or five years old, but died a year later. Most prognoses tend to be fifteen to twenty years out from diagnosis, though it's unclear how old Walt's father was at the time and just when exactly he was diagnosed.

Is this a real, heartfelt moment between father and son, a last-ditch effort to make sure Walt Jr. remembers his namesake positively? Or is it another manipulative story concocted by Walter White? Either way, Cranston's performance here is a powerful one.

Inside *Breaking Bad*: It was writer Thomas Schnauz who originally suggested that Walt's dad had Huntington's disease, as revealed in the related podcast episode.[12]

Side RxN #11: Holly White Joins the Show

It would be quite the oversight on my part if I talked about pediatrics without talking about Baby Holly and her on-screen mother, Skyler White. As I've mentioned, *Breaking Bad* pays an awful lot of attention to Walt's cancer diagnosis and treatment, and the resulting medical bills that come from that, and rightly so. The costs of having a baby, however, were somewhat overlooked.

Using available 2014 data, in America, childbirth alone can cost $13,524 for delivery and care of the mother with an additional $3,660 for the newborn, on average. This is just for the actual day(s) of delivery; other tests, visits, procedures, and consultations are billed separately. Then, facility fees, hospital charges and doctor charges, plus lab tests, epidural, radiology, and medication costs also get factored in.

This is all in a best-case scenario delivery; complications increase costs, and a caesarean birth with complications can cost up to nearly $22,000 for childbirth alone. However, the legal maximum out-of-pocket limits are $7,150 per person, or $14,300 per family, and health insurance likely reduces that total, depending on plans and availability.[13]

Viewers only get a brief glimpse of Skyler's visit to her OB/GYN (who is played by Reis Myers McCormic, though her character doesn't get a name in the show, unlike Walt's doctors), but the scene demonstrates some of the important decisions that have to be made when considering childbirth options.

In Season 2, Episode 11: "Mandala," the doctor tells Skyler that her "fluids are on the low side" and she advises scheduling a caesarean section, but it's up to Skyler to decide. Skyler agrees with the doctor (even though Holly is eventually delivered vaginally). The comment about the fluids refers to amniotic fluids, essentially a life-support system for the growing baby. These are measured during an ultrasound through amniotic fluid index (AFI) or single deepest pocket (SDP) evaluations. Too little fluid, known as oligohydramnios, can be caused by birth defects related to development of the kidneys or the urinary tract or both, maternal complications, membrane or placental problems, or post-date pregnancies. Potential risks include birth defects and miscarriage/stillbirth if low fluids arise early on, as well as pre-term birth, labor complications, and slow growth rate known as intrauterine growth restriction (IUGR). If the expectant mother is close to delivery, like Skyler was, the attending physician will likely advise a scheduled delivery to assuage risks due to low fluid levels.[14]

A note on smoking while pregnant: Skyler got a *lot* of hate for her actions and behavior on *Breaking Bad*—most of it undeserved in my opinion—but a universal outcry was shared by all when she was seen smoking while pregnant in Season 2, Episode 4: "Down." Per the Centers for Disease Control and Prevention (CDC), smoking during pregnancy can cause premature birth, birth defects like cleft lip or cleft palate, and even infant death. Smoking makes it more difficult for a woman to get pregnant in the first place, and women who smoke while pregnant are more likely to have a miscarriage, have problems with the placenta, and put the baby at risk for sudden infant death syndrome (SIDS).[15] In other words, don't do it.

XIII Oncology:

Cancer and Treatment

From Dr. Donna J. Nelson:

The choice of physicist Werner Karl Heisenberg (1901–1976) as the alter ego for Walter White's evolution during *Breaking Bad* was appropriate for a number of reasons.

Walt and Heisenberg were both genius scientists. Heisenberg was a Nobel Prize recipient, and Walt believed he should have shared a Nobel Prize.

Heisenberg died of kidney and gall bladder cancer, similar to the fate that Walt believed he had been sentenced to—death by lung cancer. Skyler poses that a cause of Walt's cancer could have been due to his exposure to chemicals in a chemical application lab many years earlier, when Walt had complained about not having the right kind of ventilation hood; likewise Heisenberg had worked with uranium for years. Walt's cancer can be taken as symbolizing his decaying loss in his personal life throughout the show.

In taking the name Heisenberg, one associates the Heisenberg Uncertainty Principle with Walt. This principle holds that there is a limit to the precision with which certain pairs of physical properties of a particle can be known, such as location and momentum. The association of Walt with the Uncertainty Principle is appropriate because Walt's foes cannot determine simultaneously his location and his next moves.

The German name for Heisenberg's principle was translated as the Heisenberg Uncertainty Principle. However, in his paper on it, Heisenberg used the word "*Ungenauigkeit*" (imprecision). Parallel to this, it was Walt's concern to avoid imprecision in his synthesis of meth that enabled him to create such a pure product.

101

These doctors ... talking about surviving. One year, two years, like it's the only thing that matters. But what good is it, to just survive if I am too sick to work, to enjoy a meal, to make love? For what time I have left, I want to live in my own house. I want to sleep in my own bed. I don't wanna choke down 30 or 40 pills every single day, lose my hair, and lie around too tired to get up ... and so nauseated that I can't even move my head.—Walter White, Season 1, Episode 5: "Gray Matter"

You knew this one was coming. As much as Walter White's illegal meth operation plays into the plot of *Breaking Bad*, his terminal diagnosis with lung cancer provides the ultimate springboard into the criminal world, as well as an excuse behind Walt's every decision. That's a brilliant if simple stroke of storytelling: Cancer is an infamous disease that has touched seemingly every life on Earth, either directly or indirectly. It's an instantly recognizable antagonist, just as villainous as high school bullies or goose-stepping Nazis. It's an instant sympathy-earner in dramas because the character is given a badge of courage or the mark of a survivor, or both, with one simple word: cancer.

Walter White wrings every possible emotion out of that one word. Much like our earlier discussions of chemical reactions and the transformative power of chemistry in life itself, the cancer diagnosis acts as a sort of catalyst for Walt's own transformation. Before it, he was simply existing; after coming to grips with it and making the insane decision to produce illegal methamphetamine in order to provide for his family, Walt truly starts living even as his body itself is rapidly shutting down. Cancer and the ever-present specter of death loom over Walt at every turn—a fact he uses to his advantage to gain sympathy quite often—even as he dodges more violent and instant causes of death with little more than his scientific acumen and the devil's own luck.

A significant chunk of screen time is given to Walt in various stages of his battle with cancer, from the diagnosis very early on in the series, to the trying times of his chemotherapy and surgery, to his remission, and ultimately to his cancer's return. In just about two years of TV time, we go from seeing Walt try (briefly) to handle the cancer diagnosis himself, to seeing him share the awful news with his family who instantly start arguing amongst themselves over the best course of action for him—treatment or no treatment—to various methods of treatment and hospital visits that exacerbate Walt's symptoms; these story elements are adapted from the real-life experiences of millions of people afflicted with the disease.

Cancer is far too big a topic to tackle in a chapter, but essentially, it's a group of diseases in which abnormal cell growth occurs, with the potential for this growth to spread throughout the body. I'll be focusing this discussion on lung cancer—uncontrolled cell growth in the lung tissue—specifically since that's the particular disease Walt deals with, though it goes without saying that cancer can affect any of the body's cells, tissues, organs, and organ systems. Its ubiquity and variety helps to explain why this is a trillion-dollar-plus medical cost annually, worldwide.[1]

Walt's cancer is an ever-present thread throughout *Breaking Bad*, even when it takes a bit of a back seat to the more thrilling and less realistic arc of becoming a meth-dealing kingpin, but to understand how well the show handled it, we'll have to get a little more advanced.

Advanced

What Is Lung Cancer?

Walt, a nonsmoker who just turned fifty the day before he's diagnosed with inoperable lung cancer by Dr. Belknap, is given a couple of years to live (at most) even with chemotherapy. This is the inciting incident for Walt's transformation into Heisenberg, but it's also an enemy that Walt cannot outsmart, outgun, or bargain with, so cancer continues to rear its ugly head throughout the series. It's not until Walt decides to pursue treatment, thanks in part to encouragement from his family, that we get a few more specifics on his diagnosis.

It's the physician and "Top 10" American oncologist Dr. Delcavoli who oversees Walt's treatment throughout the first two seasons—who's also briefly flummoxed by Walt's brush with a false fugue state—and eventually sees the cancer go into remission, which is a lessening of the disease's seriousness or even a temporary recovery. It's Delcavoli who delivers the diagnosis of "non-small-cell adenocarcinoma, Stage 3A" in Season 1, Episode 4: "Cancer Man."

The "non-small-cell" part differentiates this type of cancer from, as you might have guessed, "small-cell" carcinoma, which accounts for 10–15 percent of lung cancers and is also known as oat-cell carcinoma due to the flattened shape of the cells when viewed under a microscope.[2] Non-small-cell lung cancer includes three main subtypes—adenocarcinoma, squamous cell/epidermoid, and large cell/undifferentiated—depending on the type of lung cell affected, along with less common subtypes. "Stage 3A" refers to the cancer's spread from the lung to nearby lymph nodes or other structures and organs, likely making surgical removal impossible.[3]

"Adenocarcinoma," found in outer parts of the lung, occurs in the early stages of lung cells that will eventually secrete mucus ("adeno-" means "pertaining to the gland"); this subtype accounts for 40 percent of lung cancers. While adenocarcinomas are more common in women than in men and tend to occur in younger patients than most lung cancers, it's the most common lung cancer to be seen in nonsmokers.[4] That last trait specifically applies to Walter White who is described in the show as a nonsmoker, but it's suggested that his exposure to chemicals earlier in his career might be the cause of his cancer. How accurate is this?

What Causes Lung Cancer?

The first step in understanding what causes lung cancer is identifying risk factors. These include smoking tobacco (thought to account for 80 percent of lung cancer deaths), secondhand smoke, exposure to radon (the second-leading cause of lung cancer in the United States and the leading cause among nonsmokers), and exposure to asbestos.

Other carcinogens (i.e., cancer-causing agents) include diesel exhaust, radioactive ores like uranium, and inhaled chemicals like arsenic, beryllium, cadmium, silica, vinyl chloride, nickel compounds, chromium compounds, coal products, mustard gas, and chloromethyl ethers. High levels of arsenic in drinking water, air pollution in cities, and even the dietary supplement beta-carotene have been related to an increased risk of lung cancer.[5]

Most of these risk factors are preventable and can be minimized or removed completely since they're either personal habits or hazards associated with the home or workplace. In Season 1, Episode 4: "Cancer Man," Skyler seems to think that Walt's shocking diagnosis stems from his work in what she calls the "application's lab" twenty years earlier with "all those chemicals" around, on at least one occasion, without a proper ventilation hood. Walt shakes this idea off, saying, "We always took the proper precautions."

However, there are risk factors that can't be changed. These include personal and family history with lung cancer; having lung cancer increases the risk for developing lung cancer again in the future, and genetics plays a role in cancer risk. Radiation therapy to the chest as a treatment for other cancers also increases the risk for lung cancer, especially for smokers.

What exactly caused Walt's cancer is left up to speculation. However, regardless of the specific factor, the mechanism of cancer remains the same: damage to a cell's DNA, specifically the genes that regulate cell growth and differentiation. This abnormal cell growth can form either benign, noncancerous tumors or malignant tumors that can spread throughout the body—known as metastasis—wreaking havoc on its many systems.

What Are the Symptoms?

In *Breaking Bad*, Bryan Cranston acts out both early and late-stage symptoms of his lung cancer in addition to the very real side effects to his treatment; more on those in a minute. In early stages of lung cancer, patients experience: coughing, including a persistent, chronic cough; coughing up blood or rust-colored spit and phlegm; difficulty breathing and wheezy breathing, known as "stridor"; loss of appetite leading to weight loss; fatigue; and recurring infections like bronchitis and pneumonia.[6] Walt's cancer has likely been affecting him for quite some time before the pilot of *Breaking Bad* since it's his collapse that ultimately sends him to the doctor for a diagnosis. And when it comes to showing how Walt's cancer is progressing or even returning, the show's shorthand is to have Cranston coughing up blood on a number of occasions.

The late-stage symptoms of lung cancer are a little tougher to portray on camera, especially when the main focus of the story is less about the disease and more about the disintegration of Walter White's morality. These symptoms, which occur as the cancer moves throughout the body, include: bone pain; swelling in the face, arms or neck; headaches, dizziness or weakness/numbness in the limbs; jaundice, a yellowish coloration of the skin; and lumps in the neck or collarbone region.[7] Most of these symptoms are left up to the audience to infer, though they go a long way toward explaining Walt's chronic agitation and persistent pain. If caught early enough, however, lung cancer can be treated.

How Is Lung Cancer Treated?

The first step in treatment is a proper and exhaustive diagnosis, a process that can take three to five days. The custom diagnosis is necessary for a treatment plan that works for an individual since no two patients or cancers are identical. Diagnostic evaluations include: imaging tests like x-rays and CT scans of the lungs, sputum cytology (in other words, looking at coughed-up spit and mucus under a microscope for cancer cells), and a tissue sample, or biopsy. This biopsy can be performed through a bronchoscopy (examining the lungs through a lighted tube passed down the throat), mediastinoscopy (samples of lymph nodes taken from behind the breastbone through a surgical incision), or a needle biopsy, which uses x-ray or CT scan images to guide a needle into lung tissue for collection of abnormal cells. After all of that, a doctor will determine the staging (or progression) of cancer by another series of imaging tests, including CT scans, MRI, positron emission tomography (PET), and bone scans.[8]

Depending on the diagnosis, the treatment plan will differ on a patient-by-patient basis. Some patients will opt for no cancer treatment at all, preferring "comfort care" to treat the symptoms instead of the cause, rather than go through the taxing and grueling process of surgery and various therapies. Walt's extended family has this very conversation with him in Season 1, Episode 5: "Gray Matter" in which everyone weighs in on whether or not Walt should undergo treatment. It's a tough scene, especially when Walt Jr. takes his dad to task for presumably being too afraid to go through treatment even as Walt Jr. himself struggles to overcome his cerebral palsy, asking what would have happened if Walt would have given up on his own son as quickly as he is giving up now.

Surprisingly, it's Skyler's sister Marie, a radiologic technologist who knows quite a bit about the previously mentioned imaging diagnostics and radiation therapy treatments, who supports Walt's initial decision to avoid treatment. It all comes down to Walt's thought process: Does he choose to live out his few remaining days enjoying life to the fullest, or will he undertake the pain, extraordinary cost, and debilitating side effects of treatment for a bit more time on this Earth? Walt opts not to go through with therapy at first, but ultimately, he decides to undergo treatment, though on his own terms, a strong theme that pervades the rest of the series. For Walt, it's either his way or the Heisenberg way.

But just because Walt decides to go through treatment, that doesn't mean it's going to be an easy road. According to Delcavoli, side effects of the various treatments "can be mild, to practically non-existent. Or they can be pretty darn awful." These include hair loss, fatigue, lethargy, weight loss, loss of appetite, gastrointestinal issues, muscle aches and pains, sore and bleeding gums, nausea, kidney and bladder irritation, increased susceptibility to bruising and bleeding, sexual dysfunction, dry skin ... the list goes on.

Regardless of the side effects, Dr. Delcavoli, who prefers the term "treatable" to "curable," has confidence in the proven track record of radiation and chemotherapy, which he explains in Season 1, Episode 4: "Cancer Man." Additionally, optimism and support from loved ones certainly helps patients to get through the process. As Delcavoli says in Season 1, Episode 7: "A No-Rough-Stuff-Type Deal" in response to Skyler suggesting alternative medicine such as Eastern healing and holistic approaches, "Having a better outlook can make a tremendous difference."

However, the proven, repeatable method of treatment usually includes a combination of the following procedures:

- Chemotherapy—Anti-cancer drugs delivered intravenously or orally. The combination of drugs continues in a series of treatments over weeks or months with built-in recovery breaks. This is used pre-surgery to shrink cancer cells for easier removal or post-surgery, or both, to kill off remaining cancer cells; sometimes it can be used to relieve pain or other symptoms.
- Radiation therapy—X-rays and protons directed at the affected areas of the lungs are used to kill cancer cells. These are applied either from outside the body using external beam radiation or inside the body through brachytherapy, which uses implants—needles, seeds, or catheters—to place a radiation source near the cancer itself. Stereotactic body radiotherapy uses multiple beams of radiation aimed at the cancer from different angles and may be used in place of surgery to destroy small tumors. Radiation therapy can be used pre- or post-surgery as well, along with therapeutic, pain-relieving applications.
- Surgery—Removal of a small portion (wedge resection), larger portion (segmental resection), an entire lobe (lobectomy), or an entire lung (pneumonectomy); lymph nodes from the chest may also be removed. Risks include bleeding and infection, and complications include difficulty breathing afterward until the lung tissue expands over time to make it easier to breathe.

Walt's diagnosis of an inoperable lung tumor requires him to go through chemotherapy and radiation therapy in order to attempt to shrink it down to a more manageable size. Progress of either the cancer itself or the treatment's efficacy can be measured through a variety of tests and imaging scans like the PET scan Delcavoli mentions in Season 1, Episode 7: "A No-Rough-Stuff-Type Deal" as well as either diagnostic or exploratory MRIs. In this same episode, Delcavoli also believes they've got the "antiemetics tuned right" to keep Walt from feeling nauseous, but wants to look at another PET scan following the round of chemotherapy to reevaluate.

By Season 2, Episode 5: "Breakage," Walt has completed his first round of treatment and there's reason to be optimistic; by "4 Days Out," the ninth episode of that season, Walt goes for a full PET/CT scan, and the results confirm that the tumor shrank by 80 percent. As long as the tumor hadn't grown, Walt would technically have been in remission, but this was better than even Delcavoli had hoped for, initially aiming for a 25–35 percent reduction in mass.

This comes as great news for Walt and his family, but an anomaly on the scan temporarily terrifies Walt himself since it looks like a much larger tumor. However, this block is actually lung tissue inflammation known as radiation pneumonitis, a "fairly common" reaction to his radiotherapy, which explains his cough; Delcavoli prescribes the immunosuppressive, anti-inflammatory, corticosteroid drug prednisone to take care of it. Complicating matters a bit is the fact that Walt's also coughing up blood, likely due to a tear in his esophagus from all that coughing, which Delcavoli wants to address immediately. Though Walt deserves to be chastised here—and more so considering that this was one of his rare opportunities to put a stop to his drug-dealing ways, but he chose not to take it—the good news is that the treatment has bought him more time.

In Season 2, Episode 11, "Mandala," Walt meets with his thoracic surgeon Dr. Bravenac for the first time. Despite the initial diagnosis that the tumor was inoperable, after the first round of treatment, the doctors suggest a lobectomy as a "viable," "pretty good" option. This matches up with Bravenac's self-described "pretty good" track record for performing such surgeries following full-dosage radiation treatment. It's a risky and aggressive path, but otherwise it's just a matter of time until the cancer spreads. Walt undergoes the operation itself in the Season 2 finale, "ABQ," allowing viewers to watch along with the surgeon thanks to a well-shot and beautifully edited medical montage inside the operating room. It's all scrubbing in, lots of iodine, gleaming surgical instruments, and some clinical but gory shots of the sizable lobe of Walt's afflicted lung being removed.

The beauty of montages is that audiences get to jump ahead in time and that characters get to move forward in their own story. Walt gets the results from his surgery during what appears to be the next day but is actually about six weeks later. The news is optimistic once more, though no hard figures are given. Walt continues his treatment at the oncology clinic in the eighth episode of Season 4, "Hermanos," where a conversation with another cancer patient by the name of Gary reveals that Walt's starting to blur the lines with his alter ego. Walt talks about the PET/CT scan paired with x-ray tomography as part of his radiation therapy, but also slips into Heisenberg mode for a moment when talking to the very fearful Gary:

> Oh, to hell with your cancer. I've been living with cancer for the better part of a year. Right from the start, it's a death sentence. It's what they keep telling me. Well, guess what? Every life comes with a death sentence. So every few months, I come in here for my regular scan, knowing full well that one of these times—hell, maybe even today—I'm going to hear some bad news. But until then, who's in charge? Me. That's how I live my life.

For more than a year, Walt's anxiety over his cancer takes a back seat to the concerns of his alter ego, Heisenberg, but by Season 5, Episode 8: "Gliding Over All," Walt learns that his cancer is in fact back with a vengeance.

Fun Fact: In an interesting editing decision, Walt actually got a clean bill of health in this episode by way of a voiceover scene from Dr. Delcavoli, but this scene was cut and the test results were left ambiguous.[9] This decision put the writers onto the path of Walt hiding the fact that his cancer had returned.[10] By Season 5, Episode 9: "Blood Money," it's revealed that the results weren't so great after all.

He's back on chemo at this point, but, as he tells Hank—who has just figured out that Walt is Heisenberg before throwing his laundry list of sins in his face—he's a dying man who runs a car wash, nothing more, and has less than six months to live. Despite resuming treatment in what's likely a losing battle against the second onset of cancer, it's Walt's criminal life that ultimately does him in.

Inside *Breaking Bad*: The location used for the hospital scenes for Walt's lobectomy were shot on location at Albuquerque's Gibson Medical Center, formerly the Old Lovelace Hospital (also referenced in *Better Call Saul*), as discussed by Gilligan and producer Melissa Bernstein on Episode 412 of the

Breaking Bad Insider Podcast. The defunct hospital was also used for scenes involving Walt's fugue state, Skyler's medical checkups and Holly's birth, Hank's recovery, and Brock's treatment for poisoning.

Oh, and Marie Schrader's career as a radiologic technologist? That was all actor Betsy Brandt's idea. She wanted Marie to be a medical professional, but not a doctor or nurse.[11] Radiologic technologists are medical personnel responsible for performing diagnostic imaging examinations and administering radiation therapy treatments, an occupation that crosses paths with Walter White quite a bit and gives Marie enough insider knowledge to lend weight to her advice in the show.[12]

Side RxN #12: Walt's Major Award

In the very first episode of the series, we're introduced to the genius of Walter White without any dialogue or lab-based action. We get a shot of a memorial plaque from the Los Alamos, New Mexico Science Research Center, which recognizes Walter H. White as the "Crystallography Project Leader for Proton Radiography" in 1985, and as a "Contributor to Research Awarded the Nobel Prize." It's displayed, not coincidentally, on the wall next to his Award of Merit from the New Mexico Public School System.

So not only do we learn very early on that Walt is incredibly smart and vastly overqualified to be teaching high school chemistry, but that he also has a daily reminder that his significant academic achievements have been overlooked in his adult life. That's Walter White's pathos right there in a nutshell.

As for the actual 1985 Nobel Prize in Chemistry, that was jointly awarded to winners Herbert A. Hauptman and Jerome Karle "for their outstanding achievements in the development of direct methods for the determination of crystal structures." The late Hauptman, an American mathematician, developed a pioneering mathematical model aimed at determining molecular structures of crystallized chemicals. His dual PhDs in mathematics and physics made him perfectly suited for solving an intractable x-ray crystallography problem. The late Karle, who worked on the Manhattan Project while at the University of Chicago with his wife Dr. Isabella Karle, was instrumental in achieving a three-dimensional direct method approach to determining structures of molecules up to 1,000 atoms.[13]

XIV Toxicology:

Ricin, Lily of the Valley, and ... Cyanide?

From Dr. Donna J. Nelson:

Because of my *Breaking Bad* science advising, I was able to organize two symposia for national meetings of the American Chemical Society (ACS) in 2011. These were important, because they changed attitudes about opportunities for chemists as science advisors in Hollywood, and ACS members began to think it might be possible to take on this role. It also informed scientists that some shows had science advisors and were trying to show correct science on TV. The ability to see the advisors speak and pose questions to them was a totally new experience.

The first symposium was organized for the 241st ACS National Meeting in Anaheim. Moira Walley-Beckett spoke about the goal of the *Breaking Bad* crew to get the science right. The show had long ago hit the score of 100 percent in Rotten Tomatoes, indicating its popularity with the general public. This was in March 2011, during the writing for Season 4. The show was just about to cross the 90th percentile ranking among critics (89 percent in 2010 and 96 percent in 2011), indicating its strongly positive critical reception. Due to its Metacritic score of 99 percent at the end of its five-season run, *Breaking Bad* set a record for the highest-rated TV series in history by the Guinness Book of World Records in 2014.[1]

Of course, it didn't hurt that on March 2, 2011, in addition to Moira Walley-Beckett, we had Kath Lingenfelter from the show *House MD*, Jamie Paglia from *Eureka*, Kevin Grazier from *Zula Patrol*, *Eureka*, and *Battlestar Galactica*, and science writers Sidney Perkowitz and Mark Griep. This was the first time my ACS colleagues and I had ever assembled a panel of science advisors for an ACS National Meeting symposium. When we made the commitment to attempt this, we didn't know whether we could get these speakers to attend; we hoped that the location of the meeting (Anaheim, California) would help, because it is so close to Burbank, and that turned out to be true. The symposium was immensely successful, with standing room only in the room seating about six hundred people.

Similarly, the second symposium in August 2011, at the following ACS National Meeting in Denver, featured science advisors Jane Espenson (*Caprica, Buffy the Vampire Slayer*, and *Battlestar Galactica*), Aaron Thomas (*CSI:NY*), and Corinne Marrinan (*CSI* franchise and *Code Black*), in addition to myself.

101

Walter White: Castor beans.

Jesse Pinkman: So, what are we going to do with them? Are we just gonna grow a magic beanstalk? Huh? Climb it and escape?

Walter White: We are going to process them into ricin.

Jesse Pinkman: Rice 'n Beans?

Walter White: Ricin. It's an extremely effective poison.—Season 2, Episode 1: "Seven Thirty-Seven"

Poisons have long been a favorite mode of murder for fiction writers. They're the sexy, secret substances of spycraft, supposedly undetectable and untraceable, the key component to committing the perfect crime. From Joseph Kesselring's play *Arsenic and Old Lace* (later adapted into Frank Capra's classic 1944 film of the same name) and the many and varied killer chemicals used by Agatha Christie (who gleaned her knowledge of poisons while working in hospital pharmacies during the World Wars) in her classic mysteries, to modern murder mysteries and procedural thrillers, perhaps it's no surprise that poisons made their way into a show as heady as *Breaking Bad*.

There is, however, a surprising amount of "murder by poisoning" in this show, or at least attempts to do so. The subtle-but-sinister killers are often overshadowed by more overt violence at the hands of drug dealers (or Heisenberg himself), but their psychological impact on viewers is longer lasting. *Breaking Bad* fans, including yours truly, should get goose bumps at the mere mention of the words "ricin" and "lily of the valley." (You're forgiven if you don't remember the show's brief mention of saxitoxin.)

Ricin, teased throughout the series, is the "Chekhov's Gun" of *Breaking Bad*. (The expression "Chekhov's Gun" refers to Anton Chekhov's famous storytelling advice: "If in the first act you have hung a pistol on the wall, then in the following one it should be fired. Otherwise, don't put it there.") The first appearance of ricin occurs in the Season 2 premiere, "Seven-Thirty Seven," where Walt conceives of the idea to use it in order to eliminate psychotic Mexican drug kingpin Tuco Salamanca, no muss, no fuss. This didn't go as planned. Walt found another chance to whip up a batch of ricin in Season 4, Episode 7: "Problem Dog," this time with the intention of killing Gustavo Fring. That didn't work either, but at least the ricin wasn't used to poison Brock Cantillo—the son of Jesse's girlfriend, Andrea—despite Jesse's assumption to the contrary. The potent poison comes back into play in the final season where it's a constant threat to anyone who challenges Walt. Ultimately, it's Lydia Rodarte-Quayle, Gus Fring's methylamine supplier, who is the sole character to die from ricin poisoning in the series finale, "Felina."

Ricin itself is made in the seeds, or "beans" (they're not true beans) of the castor oil plant. Poisonous through inhalation, injection, ingestion, or absorption, ricin messes with the body's protein-making machinery and shuts down cells' most basic functions. Depending on the dose of ricin and how it enters the body, symptoms will differ and can take hours or days to appear, an aspect that makes it appealing for fictional poisoners who need to stymie investigators and establish alibies. Symptoms following ingestion include severe nausea, vomiting, diarrhea, and difficulty swallowing, followed by bloody feces and vomiting blood; inhalation and absorption symptoms range from a cough and fever to symptoms resembling a severe allergy. Symptoms can persist for up to a week, but ultimately, without treatment to minimize the poison's effects, organ failure and effects on the central nervous system can lead to death.

Lily of the valley, meanwhile, needs no modification to be poisonous. This unassuming woodland plant is nearly as notorious as ricin as far as *Breaking Bad* fans are concerned. Sometime during the events of Episode 12, Season 4: "End Times," Walt (or someone acting on Walt's orders) poisons Brock with the lily of the valley plant, which the child's doctor later confirms. This is done off-screen, but the Season 4 finale "Face-Off" closes on a shot of the plant in Walt's backyard patio, strongly implying his role in the poisoning of the boy. Jesse figures it out for himself thanks to Saul's involvement in the aptly titled Episode 11, Season 5: "Confessions." Walt himself cops to the meticulously planned poisoning in Season 5, Episode 13: "To'hajiilee."

The brightly colored berries of this sweet-smelling flowering plant may draw the most attention—especially for children—but the stems, leaves, and flowers are also poisonous if ingested. The plant contains more than three-dozen chemicals that can affect the heart, and even small amounts can cause abdominal pain and vomiting, a reduction in heart rate, blurred vision, drowsiness, and rashes. Children and pets are particularly susceptible since they're more likely to eat the plant, but treatment is possible.

Oh, and saxitoxin? This one probably won't immediately come to mind since it was only mentioned in passing by Marie during Season 5, Episode 12: "Rabid Dog" as a potential method for permanently taking out Walt, whom she and Hank know to be Heisenberg at this time. While Marie is in no danger of breaking bad here—beyond a little shoplifting now and then—saxitoxin is no joke. Let's just say that the chemical's potential for paralysis might have you second-guessing that extra order of shellfish since the nasty neurotoxin made by some marine dinoflagellates and freshwater cyanobacteria accumulates in these otherwise tasty treats.

To understand how all of these poisons affect the human body and to see how close *Breaking Bad* actually came to reality, we'll have to get a little more advanced.

> Jesse Pinkman: He's not sick, he was poisoned.
> Gus Fring: How did that happen?
> Jesse Pinkman: The doctors, they don't know.—Season 4, Episode 12: "End Times"

Advanced

What Is Toxicology?

Toxicology is a multidisciplinary study of the adverse effects of chemical substances on living things, along with the diagnosis and treatment of exposure to said substances. These chemicals include toxins, which are produced within living cells or organisms, and toxicants, which are artificially produced. The area of study overlaps with biology, chemistry, pharmacology, medicine, and nursing, pulling from a lot of disciplines for a very specific purpose. The threat of poisoning is not only a fantastic way of adding drama, intrigue, and tension to the plot of *Breaking Bad*, it's also a perfect cross-section of the many scientific disciplines on display.

While ricin tends to get the lion's share of the attention on the show, there are other poisons referenced and used throughout *Breaking Bad*. Let's look at each of them in turn:

Saxitoxin

This potent neurotoxin accumulates in shellfish like scallops, clams, oysters, and mussels; it's also been found in puffer fish and tilapia. Saxitoxin is water-insoluble, heat-stable, acid-stable, and can be stored in shellfish for several weeks or up to two years. The toxins are not eliminated by ordinary cooking methods. In humans, it's the toxin responsible for the illness known as "paralytic shellfish poisoning" or PSP.

Saxitoxin works as a neurotoxin that blocks sodium channels of neurons, which means that it prevents normal cellular function, leading to paralysis. Basically, if these cells aren't allowed to function normally, communication with the afflicted region of the body becomes impossible and paralysis occurs. While this gives the outward appearance of leaving the victim calm as the PSP symptoms progress, they're still conscious right up until death from respiratory failure. Its median LD_{50}, shorthand for "lethal dose, 50%," or a given dose required to kill half of a test population after a certain period of time, is only about 5.7μg/kg (micrograms per kilogram), or 0.57mg for a 100kg (kilogram) individual, if ingested; for perspective, this amount is smaller than a grain of sand. The lethal dose is about ten times lower (50μg for a 100kg individual) for saxitoxin entering the body through a wound or injection. When aerosolized, saxitoxin toxicity is around 5mg·min/m^3 (milligram-minutes per cubic meter), a measurement that takes variable breathing rates into account. (For example, breathing in one milligram of saxitoxin per cubic meter of air for one minute equals 1mg·min/m^3; breathing in two milligrams of saxitoxin per cubic meter of air for thirty seconds, or half a milligram of saxitoxin per cubic meter of air for two minutes, also equals 1mg·min/m^3.) This extremely low value made saxitoxin a worthwhile candidate for weaponization by world militaries; it's listed under Schedule 1 of the Chemical Weapons Convention treaty, meaning it has few if any uses outside of chemical weaponry.

Treatment includes gastrointestinal decontamination with activated charcoal and gastric lavage (stomach pumping or gastric irrigation) with an alkaline solution since it reduces the toxin's potency, along with around-the-clock monitoring and airway management to deal with respiratory paralysis. Drugs used to improve muscle weakness have yet to go through clinical trials for evaluation, however.[2] Luckily, no one on the show had to be a guinea pig (which are, ironically, even more susceptible to the toxin than humans) since the highly lethal saxitoxin was only mentioned once. To quote Hank: "Jesus, Marie!"

Lily of the Valley

This lovely-sounding flower is quite common with varieties arising in China, Japan, Eurasia, and the United States. Every part of *Convallaria majalis* is poisonous, from its bright orange-red berries and sweet-smelling flowers to its spreading, underground stems (or rhizomes) and leaves. This toxicity comes as part of the plant's natural defense systems that evolved to keep animals from eating it. The specific poisonous power of the lily of the valley is due to the plant's thirty-eight or so cardiac glycosides, which are organic compounds that inhibit the heart muscle cell's sodium-potassium adenosine triphosphatase (Na^+/K^+-ATPase) pump. This important cell membrane enzyme is vital to cell physiology, so when the glycosides alter the pump's function, heart rate decreases and both the force of the contractions and the volume of blood pumped increase. This sounds bad, but it can actually be therapeutic in treating arrhythmia and congestive heart failure, historically; synthetic drugs have replaced the naturally occurring chemicals due to toxicity concerns and to better control dosages.

However, the glycosides found in lily of the valley plants can do a fair amount of damage in other areas of the body besides the heart. The main culprits are the chemicals convallarin, convallamarin, and convallatoxin. In addition to the circulatory system effects, symptoms of lily of the valley poisoning can include afflictions of the eyes, ears, nose, mouth, and throat (blurred vision, and yellow, green, or white halos around the eyes*); the gastrointestinal system (diarrhea, loss of appetite,* stomach pain, vomiting or nausea); the skin (a rash or hives); and the nervous system (confusion, depression,* disorientation, drowsiness, fainting, headache, lethargy, and weakness). *These symptoms are seen only in chronic overdose cases.

Suspected poisoning should be followed up with immediate medical help, which is exactly what Andrea Cantillo did by taking Brock to the hospital when he fell ill. In the scene, Andrea describes his condition as being like a flu that keeps getting worse, which jibes with many of the symptoms listed above; Brock was ultimately admitted into the pediatric ICU for care. Treatment includes the monitoring of vital signs, consumption of activated charcoal, assisted breathing, and intravenous fluids; severe cases may require use of an electrocardiogram to monitor heartbeat and a temporary pacemaker.[3] All of this is to say that Brock's physical response to being poisoned with lily of the valley seems accurate; it's actually Walt's assertion that he planned out exactly how much to give the boy that's a bit specious due to all the factors involved.

Ricin

Ricinus communis, also known as the castor bean or castor oil plant, is a perennial flowering plant found in the southeastern Mediterranean region, Eastern Africa, and India, and is the only species in its genus. The plant's seed, the castor bean, is comprised of about 50 percent castor oil and the water-soluble toxin known as ricin. This highly toxic chemical is a lectin, a protein that binds carbohydrates. Even more specifically, it's classified as a Type 2 ribosome-inactivating protein, a mouthful of a name that's helpfully but ironically shortened to RIP. The two protein chains present in this chemical work together to invade and alter cells, inhibiting protein synthesis by preventing the ribosome's messenger RNA from assembling amino acids. Without this essential function of cell growth and maintenance, cells, tissues, and systems soon start to break down. RIP, indeed.

If ricin is ingested, say by unknowingly pouring a packet of it into your chamomile tea, the effects could show up in as little as six hours.

Fun Fact: Ricin, a protein, can be denatured at temperatures above 80°C/175°F, which is right in the range of the temperature at which hot beverages are served. Walt got lucky on this one.[4]

While it's possible that the ricin could be degraded by enzymes in the digestive system, any surviving ricin could still cause injuries to the mucous membranes of the gastrointestinal tract.

Within a matter of hours of ingestion, or possibly up to five days after, pain, inflammation, and hemorrhaging of the GI tract's mucous membranes can occur. These symptoms progress in severity to the point that a patient experiences bloody vomit and stool. Due to all this fluid loss, low blood volume can lead to pancreas, kidney, and liver failure, followed by shock, indicated by disorientation and stupor, weakness and drowsiness, and excessive thirst along with low urine production and blood in the urine. Despite all of these nasty symptoms, ingesting castor plant seeds prevents much of the ricin from entering the body's systems due to the seeds' indigestible coat. However, thoroughly chewing the beans and chowing down on more than a half-dozen or so of them can prove fatal to an adult since the pulp contains more ricin ... which is exactly why Walt processes out the oil and concentrates the leftover toxin.

A quick montage in the Season 2 premiere, "Seven Thirty-Seven," shows Walt and Jesse making ricin from castor beans, which makes this a great time to remind you to **never try this at home**. Ricin is monitored by the U.S. Department of Health and Human Services and is classified as an extremely hazardous substance, one that's subject to strict requirements by production, research, and storage facilities. But purely for intellectual curiosity, Walt's process of removing the beans' outer coats, cooking them, mashing and filtering the pulp, and using solvents to extract the ricin is essentially spot-on.[5]

Fun Fact: In the ricin-making montage, peanuts were used to stand in for the castor beans.[6]

If Walt's ricin was as pure as his methamphetamine was known to be, he would have needed about a grain of salt's worth of the toxin to take out Tuco, Gus, or Lydia. And since there's no proven antidote available, once it's been delivered, the deed is all but done; death can take place between thirty-six and seventy-two hours after exposure.[7]

This all sounds rather nasty, but with sufficient treatment most patients will fully recover. The CDC suggests avoiding ricin exposure in the first place, which sounds like good advice. However, if you happen to run up against Heisenberg and suspect that you've been poisoned by ricin, the two main approaches to treatment are to rid the body of ricin as quickly as possible and to provide support related to the symptoms. Quick removal and disposal of clothes, along with washing yourself, is the first step in minimizing ricin's effects. Depending on the manner of exposure, treatment includes breathing support, IV fluids, medications to treat seizures and low blood pressure, flushing the stomach with activated charcoal, and washing out the eyes. Counterintuitively, neither induced vomiting nor consumption of fluids should be recommended to someone thought to have ingested ricin.[8]

While it's possible that Lydia Rodarte-Quayle is actually still alive out there in the world somewhere—especially since Walt gave her the exact manner of her poisoning and she had been exposed to the ricin about twelve hours earlier, giving her plenty of time to seek medical attention—there's one more case of poisoning that has a more definitive ending. This one uses a mystery poison that the *Breaking Bad* creative team isn't divulging, but one that was researched extensively by writers assistant Gordon Smith.[9]

Mystery Poison

Just what, exactly, the offending chemical is supposed to be, we may never know, but it was used by Gus Fring in Season 4, Episode 10: "Salud" to poison the entire Mexican cartel leadership by spiking the (fictional) tequila, Zafiro Añejo. We see Fring consume the poisoned potable as well, but he's also seen inducing vomiting—**which you should never do**; it's counterintuitive but only a small amount of the poison actually is expelled in vomiting while the rest is driven deeper into the digestive system.

Additionally, in the related *Breaking Bad Insider Podcast* episode, Gilligan says that Fring takes activated charcoal tablets to slow the effects of the poison.[10] This allows him the time to get the medical attention he needs to save his life as the cartel leadership dies in a fairly rapid and unpleasant manner. For my money, I'm betting Gus used cyanide since it's a fast-acting poison that can cause dizziness and loss of consciousness within a few minutes, followed by cardiac arrest, but we may never know for sure.

If you ever have questions about poisoning or poison prevention, even if it's not an emergency, you can call the National Poison Control Center at 1-800-222-1222, twenty-four hours a day, seven days a week, or visit www.poison.org.

XV Pharmacology: Drugs, Addiction, and Overdoses

From Dr. Donna J. Nelson:

From the beginning I realized that the primary goal of the writers and crew each season was to get picked up again the next season. They wanted a hit. I discussed earlier that I had aligned my goals with theirs, because I knew if I didn't, I would politely be told, "Thanks, and if we need you again, we'll let you know." And then I would not hear from them again.

However, I thought I might be able to get one message across, so long as it was aligned with their goals. I decided upon the most important message to give to this group, which would not interfere with their ability to have a hit. This message was that the public didn't appreciate science or scientists as much as they should.

In recent years, there has been a great effort on the part of foundations, societies, government, universities, schools, the media, and others to increase the general public's knowledge about science. However, there has not been a similar effort to increase the public's *appreciation* for science *and scientists*. There is a subtle but real difference.

Science is responsible for every new product that benefits our lives; science gives us our beautiful fabrics, car parts, pharmaceuticals, computer and other device components, medicinal instruments, airplanes, carpets, ceiling tiles, wall paint, and on and on. And this fabulous science is carried out by scientists.

After I decided this would be my one message that I tried to get across—the one social change I would attempt—I made this case to Vince, the directors, the writers, the actors, and everyone connected with the show with whom I came into contact. I finally realized I had made an impression when I received script pages to review for an upcoming scene in Season 4, Episode 1: "Box Cutter."

Here's a brief recap of the pertinent scene: Walt and Jesse are held in the Superlab, waiting on Gus's reaction to Gale's murder. Gus arrives, sees Victor is cooking, and begins changing into a lab suit. Walt expects that Gus is about to exact some severe consequences for the murder. He understands that he needs to talk Gus out of this. Walt also realizes that he and Victor will not be able to work together. He realizes that he can frame this situation his way, by making the case that he and Jesse are too important to be discarded. He gambles that Gus is sufficiently absorbed with the success of his meth trade that his decisions will be made purely as business decisions.

Walt presents this argument—he and Jesse are critical to the success of Gus's business; the business will succeed with Walt and Jesse, but not with Victor, because Walt and Jesse are excellent chemists. He relies on Gus's appreciation for them as chemists and their capability to do excellent chemistry. Gus buys Walt's argument, slits Victor's throat with a box cutter, then changes back to his work clothes, and tells Walt and Jesse to get back to work.

When I read this scene, I felt I had succeeded in my one goal. This scene demonstrates the importance of knowing chemistry, and of respecting the talented chemists who do it right. The scene is scientifically credible. It uses science to create drama. It presents to the public several high-level topics from chemistry. And the sharp dialogue awakens curiosity in viewers, who might try to learn about and understand the chemistry.

101

Well, then why should we do anything more than once? Should I just smoke this one cigarette? Maybe we should only have sex once, if it's the same thing. Should we just watch one sunset? Or live just one day? It's new every time, each time is a different experience.—Jane Margolis, Season 3, Episode 11: "Abiquiu"

It goes without saying that *Breaking Bad* serves as a cautionary tale against making, dealing, or using illicit substances, but this is as good a place as any to say it just the same. As much as fans of the show like to hold Walter White up as an ideal antihero, he unquestionably leaves death and destruction in his wake. His many transgressions—which surprisingly never include drug use, save for a little bit of marijuana—include numerous deaths at his hands, by his actions, and on his orders. You can measure a viewer's willingness to side with Walt based on the point at which they turn against him: Is it when he strangled Krazy-8? When he poisoned a child to manipulate Jesse? Or when he refused to act when Jane Margolis was asphyxiating?

That latter death—of which fans will argue over for years to come, as to how much blame can be laid on Walt—isn't only sad within the narrative of *Breaking Bad*, but within the broader topic of drug addiction and overdoses. The show rarely glorifies drug use or the money earned from dealing, but it does take pains to portray the dead-end, dangerous, and deadly side of drugs. Whether it's checking in on Jesse's pals and dealers Skinny Pete, Combo, and Badger; seeing the daily life of Wendy the meth-addicted street prostitute; witnessing the squalor that Spooge (the junkie who stole from Jesse), his lady, and their son live in; or the violence of Jack Welker's White Supremacist Gang, the drug culture of *Breaking Bad* becomes less and less attractive as the seasons go by. The only "winners"—temporarily anyway—are the corporate executives who oversee the drug empire but distance themselves from it, and the few drug addicts who've managed to escape their habits and maintain their sobriety.

It's important to understand just how the drugs displayed on *Breaking Bad* affect people in the real world. Specifically, it's important to understand what crystal meth is, why it's addictive, and why it's destructive. Having this knowledge at hand not only helps to heighten the experience of watching the show, but also to temper any dark desires to follow in the fictional footsteps of Walter White.

So what exactly is a drug? Basically, a drug is any non-nutritional substance that causes a change in the body when inhaled, ingested, absorbed, or otherwise consumed. Drugs include medications and pharmaceutical substances used for the treatment of disease, psychoactive chemicals used to alter mood or perception, and recreational drugs used to alter consciousness. All of these drugs—from aspirin to oxycodone, caffeine to cocaine, and marijuana to methamphetamine—fall under various regulations and classifications depending on the individual drug's chemical structure, potency, addictive properties, and side effects.

Drugs can be addictive when their actions are both rewarding and reinforcing, essentially hijacking the brain's normal reward system and eventually raising the drug's importance above all else. Addiction may start as a lifestyle choice or as a result of genetic, environmental, or psychological factors, or a combination of these, but chronic addiction actually alters physical processes in the body. These structural changes make the addiction harder to shake while also increasing the body's tolerance to the drug, requiring larger doses over time. Risks of withdrawal, overdosing, and escalation to stronger drugs further complicate matters. It's a vicious cycle, but addiction can be overcome through a combination of medication and behavioral therapy, with research stressing the importance of integrating both approaches.[1] If Jesse and Jane had only taken this treatment more seriously, the show would have gone in a very different direction.

To understand meth, its effects, and just how difficult it is for people to overcome their addiction, we'll have to get a little more advanced in our discussion.

Advanced

Understanding the mechanisms of drugs and how they affect the body is a difficult challenge that requires a multidisciplinary approach, reflected in the complex discipline known as pharmacology. This branch of biology examines the interactions between the body and the chemical substances that affect its functions. The wide-ranging field of pharmacology includes study of the make-up of a drug, its engineering and production, the mechanism of its actions and those of the body's organs and systems, cellular communications and interactions, medicinal applications, and toxicology, to name just a few. The specialization pulls from biochemistry, cell biology, physiology, genetics, neuroscience, and microbiology, which goes a long way toward illuminating just how involved and intricate the field is. Research in pharmacology leads to both new drug discoveries as well as a better understanding of the human body's inner workings, progress that the field has been undertaking since its start in the nineteenth century.

One drug that also has its roots in the 1800s is amphetamine and its derivations, methamphetamine and methamphetamine hydrochloride. Romanian chemist Lazār Edeleanu synthesized amphetamine ($C_9H_{13}N$)—which he named phenylisopropylamine before its more common name, alphamethylphenethylamine, was contracted—in Germany in 1887. A few short years later, in 1893, Japanese chemist Nagai Nagayoshi synthesized methamphetamine—N-methylamphetamine ($C_{10}H_{15}N$)—from the medicinal stimulant known as ephedrine ($C_{10}H_{15}NO$). In 1919, pharmacologist Akira Ogata synthesized methamphetamine hydrochloride ($C_{10}H_{15}N\text{-}HCl$) through reduction of ephedrine with red phosphorus and iodine. A lot of that terminology might still sound like gibberish, even at this point, but some of those words should sound familiar to *Breaking Bad* fans. I'll get more into the finer points of meth making in an upcoming chapter.

For as much as meth is a part of *Breaking Bad*, the show spends relatively little time talking about what the drug actually does to people. Perhaps it's due to the storytelling rule of "show, don't tell" and the fact that the show's visual representations of drug use are much more entertaining than reading a drug's safety data sheet. Still, it's probably best to keep the show's gorgeous cinematography and visual representations of drug use in mind as I dive into the gritty details of meth in action.

Meth is a strong stimulant—a drug that increases activity in the body, is invigorating and pleasurable, and known as an "upper." It affects the central nervous system, or brain and spinal cord. Mainly used recreationally, and less commonly as a treatment for attention deficit hyperactivity disorder (ADHD) and obesity, low doses of meth can improve mood, heighten awareness, increase concentration and energy, and reduce appetite. Higher doses, however, can result in a disconnection from reality known as psychosis, a breakdown of skeletal muscle (rhabdomyolysis), seizures, and brain bleeds, with chronic high-dosage exposure producing wild mood swings, delusions, and violent behavior.[2] (Note that patients experiencing fully developed psychosis can experience auditory and visual hallucinations when agitated, perhaps similar to Season 1, Episode 4: "Cancer Man" when Jesse hallucinated a pair of machete-and-grenade-wielding bikers coming to get him while under the effects of his meth use.[3])

Because of the low-dose effects of meth, it was used in World War II by both German and Japanese forces to keep soldiers awake and alert, and increase the productivity of industry workers; side effects and potential for abuse eventually reduced usage in this manner. There's a laundry list of side effects, from loss of appetite, to hyperactivity, to twitching and tremors, but the most recognizable effects include excoriation disorder—abnormal scratching and picking at the skin—as well as dry mouth and teeth grinding that lead to "meth teeth" or "meth mouth."[4] These very noticeable traits are best displayed on the show by the Spooges and Wendy the prostitute, and in the drug den scene in the Season 2 finale "ABQ," from which Walt rescues Jesse. This episode is perfectly placed in time between Jane's death from overdose and Jesse's first visit to a detox clinic and group therapy.

Meth is also used recreationally due to its euphoric and aphrodisiac qualities, which allow days-long sex binges as part of a "party and play" subculture. The come-down from such experiences often has severe repercussions that include hypersomnia, or excessive daytime sleepiness.[5] (Jesse uses this knowledge to provide an alibi for himself—thanks to the testimony of Wendy—in Season 2, Episode 3: "Bit by a Dead Bee.") These aspects, along with meth's high potential for addiction and dependence, and its neurotoxic effects that alter brain structure and function, strongly suggest that the drug's downsides far outweigh its benefits. This is why it's listed as a Schedule II drug—a drug with known potential for abuse and dependence, but also accepted medical use with restrictions—in the United States; marijuana remains a Schedule I drug as of this writing.

In a perfect world, all of the side effects and legal concerns related to meth use would serve as enough of a deterrent to prevent recreational use. But this is not a perfect world and neither is the one portrayed in *Breaking Bad*. The fact remains that the best way to avoid all the problems that meth brings is to never use it, as the famous campaign (and eventual internet meme) says, "Meth: Not Even Once." But since meth addiction is a real concern, and one that's portrayed on *Breaking Bad*, it's worth exploring what exactly addiction is and what treatment options are available to overcome it.

Addiction from chronic drug use can be understood and explained through neurological models since expressions of genes in certain parts of the brain are altered by regular exposure to these chemicals. Specifically, these changes occur in an area of the brain known as the nucleus accumbens, a region that plays a role in the reward and reinforcement pathways of the brain and, of course, addiction. A discussion of the individual transcription factors (proteins that bind DNA and alter gene expression) and neurotransmitters (chemical messengers that transmit signals across cells' synapses) involved in inducing an addictive state in the brain is well beyond the scope of even this advanced section, but it's fascinating to look at the intricate interplay of chemicals on the subcellular level to see how changes to those pathways can have devastating results when it comes to addiction.

In basic terms, an addiction starts to form when chronic use of stimulants like methamphetamine increases concentrations of the neurotransmitter dopamine upstream in a neural pathway. This allows for an accumulation of certain transcription factors within a neuron that physically change the structure of the brain, "rewiring" its natural function in a way that is termed "negative plasticity." This maladaptive change leads to an addictive state, one that's much more physical than many people realize and not simply solved by "mind over matter," "quitting cold turkey," or "toughening up." This model of addiction isn't restricted just to meth, but also pertains to alcohol, cannabinoids, cocaine, nicotine, and opioids, among other substances.

With chronic use of meth comes addiction, and with addiction comes increased use of meth; complicating this vicious cycle is the body's natural tendency to develop a tolerance to such drugs. The withdrawal symptoms from meth use, which include more severe and longer-lasting depression than that seen in cocaine users, show positive correlation with the level of drug tolerance. In other words, longer drug use builds tolerance and worsens withdrawal symptoms over time. In addition to health complications of meth withdrawal, or "comedown"—excessive sleepiness and lethargy, increased appetite, heightened anxiety, psychosis, paranoia, and a deep depression[6]—there's the increased risk of an overdose as tolerance rises since more meth is required to get the same high.

An overdose is the application or ingestion of a drug in greater quantities than is recommended or generally prescribed; the term does not apply to poisons, which don't have an established safe dosage. Overdoses, technically under the field of toxicology, can be acute or chronic occurrences and may result in toxicity or even death. Any drug use can lead to an overdose, though heroin and opioid overdoses are the poster drugs claiming tens of thousands of lives in the United States annually in recent years. Heroin, which is a highly addictive drug as well, carries the risk of overdose with each use since the purity, and therefore the potency, may vary. In Season 2, Jesse's paramour Jane Margolis, a recovering addict, eventually turns Jesse on to a combination of both meth and heroin, known as a "speedball." Taking these drugs together increases the potency of the high, but the stimulant (meth) may also mask the deadly effects of the depressant (heroin), making it difficult to acknowledge tolerance against potential overdose. It was heroin, however, that led to Jane's death in Season 2, Episode 12: "Phoenix."

It's a bitter pill that Jane, who was in recovery and had been clean for over a year when she met Jesse, dies from an overdose in this way, but it's an unfortunate reality of the real world as well. Recovering addicts who find themselves back in familiar environments can fall back into old habits even a year or more after their last drug use. Some former users intentionally overdose as a way out of a perceived "no way out" situation, while others, like Jane, accidentally overdose due to decreased tolerance from being without the drug for so long.[7]

This is all very dour, and it should be when considering how serious drug use can be, so I'd like to close out this section on a more hopeful note. Those same recovering drug addicts who were studied during relapse were found to have more positive outcomes if they were in an environment that was less conducive to self-destructive tendencies while also being surrounded by a support team. Jane, unfortunately, left that behind when she fell in with Jesse and fell back into those old habits, despite her father Donald's best efforts.

Walt, whose inaction certainly didn't save Jane's life, does his part not only to keep Jesse alive, but also to get him clean again. In the episode following Jane's death, "ABQ," Walt rescues Jesse from a drug den known locally as "The Shooting Gallery" and takes him to a pricey rehabilitation facility called "Serenity." Jesse seems to be doing well after leaving rehab clean and sober, even attending regular Narcotics Anonymous meetings in Season 3. While Jesse unfortunately chooses to use these meetings as a covert way of finding other meth addicts to sell his product to, if he had just stuck with the program he could have turned his life around. That would have been best for Jesse, but not exciting TV for fans of *Breaking Bad*.

Drug rehabilitation facilities also exist in the real world, of course. They seek to provide addicts with an environment conducive to safely confronting their substance abuse in an effort to put a stop to it and avoid the many complications that come with drug use. These facilities include medications to deal with depression and other symptoms related to drug use and withdrawal, as well as expert counseling and group counseling sessions. Treatment types can range from in-patient to out-patient, localized support groups, and on-call counseling, among others. And since Narcotics Anonymous (NA) is an actual organization that *Breaking Bad* calls out by name, Gilligan and writer John Shiban revealed that they reached out to NA in order to incorporate their literature, logo, and even advice from a consultant into the show.[8] You can find more about their work at NA.org.

Chemistry III

We've exhausted just about every major instance of physics and biology featured in *Breaking Bad*, but there's still more chemistry to cover. Some of you may be wondering just when I'll be getting to the meth-making discussion; sit tight, it's right around the corner. But much like you can't make meth without the right chemicals, I'll first be taking a look back at Walt and Jesse's biggest precursor boost in the entire series to see if their methylamine math checks out. And since you wouldn't know how pure Heisenberg and his competitors' products really were without a way to check them, I'll also revisit the show's various analytical methods to make sure they measure up.

In this section, I'll first walk through the impressive and ambitious train heist that lands Walt and Jesse a (fraction of a) trainload of methylamine to keep their meth-making business going in the "Dead Freight" episode of Season 5. While the logistics of robbing a train are worked out in entertaining fashion on the show, it's the impressive conversation about the rather mundane topic of dilution that I'll be focusing on.

And while you might think that the topic of analytical chemistry sounds about as dry as the Chihuahuan desert, you'd be surprised to learn that *Breaking Bad* featured a bunch of interesting equipment and techniques over the years. I'll cover them each in turn before closing out this section with the granddaddy of all chapters, a look at the meth-making science at work throughout all five seasons of *Breaking Bad*. Get ready for a lot of meth math!

XVI Methylamine:

The Solution Is Dilution

From Dr. Donna J. Nelson:

Breaking Bad frequently shot scenes on location around Albuquerque, and I was invited to go along on these. At times, it was surprisingly more complicated than shooting in a studio. In order to secure the location, streets had to be closed off, a business would be shut down for one or multiple days, or a house could be rented for days, weeks, or months. It could be a complicated activity. In some cases, the activity could have safety issues, which meant that anyone who was not being filmed needed to stay far away from the action.

There were many scenes in the desert just outside Albuquerque. One is particularly memorable because it demonstrated the entire crew's agility. Walt is beaten and falls on the ground. In one of the takes, Brian Cranston notices an ant crawling around in the desert. Rather than moving the ant to be in the scene, they estimate where the ant is going and move to that spot, in order to get the ant in the scene. I doubt that the ant ever knew anything unusual had taken place. The crew got along fabulously.

I recall how on the day that scene was shot, the wind blew somewhat harshly across the relatively flat desert. It blew so much that at times it was difficult to talk without getting sand in your mouth. Sometimes the crew had to put on more clothing to protect against the wind and sand.

In another scene filmed that day, Walt is driving through Albuquerque, but the car carrying Walt is actually mounted on a truck that also transported the cameras and cameramen. It appeared so simple on TV, but so much effort went into this shot.

The car crash scene had to be done in one take, with a stunt double inside during the actual crash. The need to get it in one take was due to both safety reasons and budgetary reasons. The crash was intentionally done in a dramatic manner so that the car spun around, throwing around its occupant with potential for injury. Also, it is expensive to smash up a perfectly good car. Bryan was filmed in the car before and after the actual crash. The shot taken from above was filmed from a camera in a crane. In this scene, I stayed almost a block away during the crash along with the rest of the crew, for obvious safety reasons.

101

We are stealing a thousand gallons of methylamine. One gallon of forty-percent aqueous methylamine solution will yield seven-point-four pounds of product. Times a thousand gallons, at forty thousand dollars per pound, comes to $296 million.—Walter White, Season 5, Episode 5: "Dead Freight," Deleted Scene

"Dilution is the solution to pollution." This was a common phrase that used to be heard in the lab—not one that I've ever been a part of, of course—but it's an old adage that has fallen out of favor and was never really correct in the first place. Normally, this phrase applies to the problem of trying to get rid of a pollutant or contaminant by reducing its concentration (i.e., diluting it) by adding lots and lots of water. In a lab setting, these dilution "solutions" occur on a relatively small scale and are also managed by waste treatment services, but in the wider world, pollution remains a major problem. Environmentalists rightly point out that "mixing zones" still legally exist as a loophole for industrial polluters to dilute just about any discharge into a waterway without applying water quality standards.[1] Accumulations of pollutants like toxins, chemicals, and bacteria can build up in the waterways themselves and in the food chain over time, so no, dilution is not the solution to pollution.

However, dilution *is* the solution to Walt's particular problem in Season 5, Episode 5: "Dead Freight." In what becomes perhaps the most extreme lesson in dilution ever told, the *Breaking Bad* writers find a novel way to get Walt and his meth-making team all the precursor he could ever need; all they have to do is rob a train. When their chemical supply, provided by Lydia Rodarte-Quayle, comes under control of the DEA, Lydia advises them about an "ocean of methylamine" ripe for the taking on a weekly freight train. (I'll get to how methylamine is used in meth making in a later chapter, along with an explanation of why Walt dismissed Mike's suggestion of going back to a pseudoephedrine cook as an alternative to the heist.) Lydia also drops specific details needed to pull off the heist successfully, which saves her life *and* her illicit business dealings.

But is Lydia's tale of a chemical-toting freight train passing through a desert dead zone actually legitimate? It probably won't surprise *Breaking Bad* fans to find out that the production team went to great lengths to make this heist as realistic as possible. Script coordinator and research assistant Jenn

Carroll did quite a bit of verification on this episode from writer-director George Mastras. They reached out to a retired hazmat train specialist to get specifics on the following details:

- Appropriate signage for chemicals and freight
- Box car IDs and the proper combination of letters and numbers
- Proper location of the tanker car, which is at least six cars back from the engine
- How far the trestle bridge had to be from the crossroads
- Verification of "Dark Territory" areas where trains lose radio and cellphone communications

That's some impressive dedication to realism! The production team used a working locomotive engine to pull a long line of actual freight cars for the scene, though production designer Mark Freeborn's crew also applied graffiti to some of the cars to give them that extra bit of believability. Even the term "Dead Freight" is a real descriptor that refers to an empty car that still has fuel and transport costs associated with it. (Perhaps planning the original location shoot in an actual "Dark Territory" wasn't the best idea though since it meant their radio and cellphone communications would have been kaput; they remedied this in time for the actual shoot.)[2] That's all well and good that the practical side of things were covered for the heist, but the show still had to get the details right when it came to the methylamine itself.

Before I get into the meth math—which Vamonos Pest employee and henchman Todd Alquist asks Walt about during the train heist, helping to clarify things for the audience—I should explain a bit about methylamine. The organic compound (CH_3NH_2) is normally sold in a solution of methanol, ethanol, or water, to name a few solvents. It's also transported as an anhydrous gas (i.e., no water present) in pressurized tanks for industrial uses, but Walt clearly states (in a deleted scene on the *Breaking Bad* Blu-ray) that they're boosting 40 percent aqueous methylamine, so we'll go with that.

Methylamine is less dense than water. Since density is a measure of mass per unit volume, for the same *volume* of each liquid, the mass of methylamine will be less than that of water. Keep in mind that an equivalent mass of water needs to be added back in to replace the mass of methylamine Walt and his crew are stealing from the tanker, because the tanker is weighed both at the beginning and end of the train's run; a discrepancy would alert the company that it had been robbed or, at the least, had a bad leak somewhere. But because of the differing densities, and specifically the differences in masses, a lesser volume of water is needed to replace the volume of stolen, less-dense methylamine.

So what about Walt's calculations and assumptions? Do they add up in the real world? Carroll also spoke to science consultant—and coauthor of this very book—Dr. Donna J. Nelson in order to verify these chemistry-related specifics. To explain how it all checks out, I'll have to get a little more advanced.

Advanced

I've dipped a toe in meth math, so let's just dive in. Walt explicitly states that the concentration of the methylamine is a 40 percent aqueous solution, but we can also verify this through calculations. The relative density of a 40 percent aqueous solution of methylamine is 0.89 (water = 1). Water has a density of 1,000kg/m^3, or kilograms per cubic meter, while 40 percent aqueous methylamine comes in at 890kg/m^3. Unfortunately, since our thieves are dealing in gallons of water and methylamine, we'll have to do some unit conversion along the way. One cubic meter is equivalent to 264.172 U.S. gallons. Let's see what we're working with:

The total volume of methylamine in the tanker comes in at 24,000 gallons (well short of the maximum capacity of a U.S. DOT-111 tanker car of around 30,000 gallons). Walt plans to take 1,000 gallons of methylamine out and replace it with the equivalent mass or "weight" of water. According to Walt, the volume of water needed for this is roughly 900.24 gallons. (Walt makes it an even 920 gallons to "account for spillage and water left in the hose.")

So for 40 percent aqueous methylamine:

1,000 U.S. gallons = 3.785 m^3, a mass of about 3,369 kg

This mass needs to be replaced with an equivalent mass of water:

3,369 kg of water is equivalent to 3.369 m^3, which equals 890 U.S. gallons

A quick way to verify this would be to use methylamine's relative density value (0.89) as compared to water. So Walt's calculations are pretty close, especially if he just did them off the top of his head, though even adjusting the density values for both water and the 40 percent aqueous methylamine for ambient temperatures doesn't quite get us to Walt's 900.24 gallons. Let's be honest, though, these precise calculations are going to be more than good enough since the heist itself will be using a powerful pump along with a crude flow meter to keep track of how many gallons are either coming in or going out; not exactly precision measurements when working under pressure in the real world.

While watching this scene, which is a fantastic one to revisit at any time, you'll notice that Walt pumps the methylamine out from the bottom of the tanker first before pumping the replacement water back into the tanker through the top. Carroll again spoke with Dr. Donna J. Nelson to figure out how quickly the denser water would sink through the other twenty-three-thousand-odd gallons of methylamine; adding the water too soon would further dilute the chemical they were trying to steal, but adding it too late would cost them precious time. Gilligan, however, isn't totally convinced that their little pump could have kept the water flowing at the same pace as the methylamine that was flowing out, but we'll give him a pass since he got his math right.

There's one more catch that Todd brings to Walt's attention as they're prepping for the heist. He understands that they have to replace the stolen methylamine with water but wisely points out that the end result is that the company will be receiving a diluted batch. Won't that put up a red flag? Walt responds that yes, indeed they'll notice the roughly 4 percent difference in concentration and will likely "blame China for sending a marginally weaker batch." But does his math check out here?

The tanker starts out with 24,000 U.S. gallons of 40 percent aqueous methylamine, which is equivalent to about 90.85m^3, totaling 80,856kg. (40 percent of that, or 32,342kg, is the methylamine itself while water accounts for the other 48,514kg.)

Walt is stealing 1,000 U.S. gallons of solution totaling 3,369kg (1,348kg of which are methylamine) and adding back 900 U.S. gallons of water, for 3,407kg.

Once the train rolls on, it now has 30,994kg of methylamine in 49,900kg of water, for a total mass of 80,894kg.

That's about a 0.05 percent difference in overall mass, which should be within the deviation limits for the train's weigh stations.

For the methylamine itself, the tanker is now a 38.33 percent solution, a "marginally weaker batch" indeed, and a difference of about 4.2 percent. Well done, Walt!

If math is not your strong suit, hold tight; we're almost done! In a deleted scene from this episode, Walt calculates that their heist will net them a record-setting amount of money that's more than history's biggest train robberies put together. Here's his breakdown:

1 gallon of 40 percent aqueous methylamine will yield them 7.4 lbs of product

7.4 lbs × 1,000 gallons, at $40,000 per pound = $296,000,000 in terms of potential equivalency

The math once again checks out, though street value for illegal meth is undoubtedly a highly volatile market. Still, not a bad day at the office.

Fun Fact: After the successful heist of 1,000 gallons of methylamine—and the unexpected murder of Drew Sharp—Mike and Jesse want out of the game. They're looking to the rival meth distributor Declan to purchase their shares of the methylamine in Season 5, Episode 6: "Buyout." Walt, obviously, does not want to give up his ill-gotten gains and has no desire to leave the "empire business." Walt's holdout leaves Mike and Jesse with exactly 666 gallons of methylamine to sell ...

XVII Let's Get Analytical

From Dr. Donna J. Nelson:

From the beginning, when I received script pages to review for accuracy, they never covered the entire episode. They were given to me on a need-to-know basis. I was surprised to learn during my first set visit in May 2011 that this was done with everyone.

After the first couple of seasons, it was necessary to take precautions against the script leaking out. The show was simply so popular, there were efforts to get the script in advance. In addition to receiving only pertinent pages, sometimes portions of those pages were redacted, by marking through lines. However, everyone understood the need, and I heard no complaints.

Regardless of the proximity of the directors to the actors, they always watched the results of the filming on their monitors, as the filming progressed. Whether the action was a few feet away in the studio or a block away as in the car crash scene, the directors watched it on these monitors. I think it gave a consistency to the filming, and it enabled them to see the footage exactly as it would appear to the viewer later on TV. Whatever the reason, it certainly worked, because *Breaking Bad* was filmed and directed fabulously.

Everyone had a chair like a director's chair. Chairs were lined up in rows behind the monitors, with the directors being closest to them. Behind them were the assistant directors and other important people. Each important person had a chair bearing their name, so they would be guaranteed a place to sit. In the next row were chairs for the actors to be filmed that day, also bearing their names. In the next row were chairs without names; this is where I sat. I noticed that the reserved status of the chairs was taken quite seriously. Typically, a person would sit in their assigned chair or they would stand. Occasionally, a person might sit for a few seconds in someone else's chair, but if the owner was seen approaching, the squatter would scamper rapidly out of the chair without being asked, due to respect.

101

I know all about your operation. My partners here tell me that you produce a meth that's 70% pure, if you're lucky. What I produce is 99.1% pure.—Walter White, Season 5, Episode 7: "Say My Name"

So just how good is Walter White when it comes to cooking meth? In the world of *Breaking Bad,* he's the best. If he wasn't, not only would he not be very interesting, but also his work wouldn't be in such high demand by every drug dealer from Albuquerque to the Czech Republic. I'll get into the science of Walt's meth cooks over the seasons in chapter 18, but for now, let's revisit the unmatchable quality, and more importantly *purity,* of Heisenberg's superior product.

In the pilot, Walt's first cook receives high praise from Jesse who calls his methamphetamine "glass grade" due to its glass-like appearance as a result of using the pseudoephedrine method. In Season 1, Episode 4: "Cancer Man," Hank and the Albuquerque branch of the DEA are just learning of a new meth cook in town, someone whose product is the purest their testing lab has ever seen: 99.1 percent pure, to be exact.

In the Season 4 premiere, "Box Cutter," the ambitious and excitable Gale Boetticher—whose immaculate notebook is certainly good laboratory practice but spells the beginning of the end for Walt and Jesse—tells his boss Gus that the purity of their competitor's meth is quite high, though he can't account for the blue color. Gale can muster meth at a purity of 96 percent, guaranteed; the competitor's sample? Ninety-nine percent, or "a touch beyond that." Gale mentions that he needs a gas chromatograph to tell for sure—more on this in a moment. (In the same episode, Boetticher is unpacking the brand-new Superlab for Gus, saying that the high price tag is worth it considering that the equipment is on the same level as that found at pharmaceutical giants like Pfizer and Merck.)

The purity of Walt's meth never really changes much throughout the series, but Jesse improves quite a bit as a cook from beginning to end. Starting out as Cap'n Cook sneaking chili powder into his "Chili P" meth, Jesse manages to pull off an impressive 96.2 percent purity when cooking his own test batch in a cartel-owned lab in Season 4, Episode 10: "Salud."

Other supporting characters who try their hand at cooking include Todd Alquist, who manages to produce meth in the range of 75–76 percent purity (with only one lab fire) by Season 5, Episode 13: "To'hajiilee." Briefly, in Season 5, Episode 10, "Buried," the cook for rival drug manufacturer Declan can only make meth that is 68 percent pure; that poor performance earns Declan's entire crew an execution order that's issued by Lydia and carried out by Jack's White Supremacist Gang. (Victor attempts his own batch but its purity is never measured since he is dispatched in a particularly gruesome manner by Gus's own hand.)

These percentages are all well and good to throw around as shorthand for letting viewers know just how far above the other cooks Walt actually is, but for a show as dedicated to scientific accuracy as *Breaking Bad*, fans demand more. Wisely, all the writers had to do was mention one piece of equipment: the gas chromatograph.

This piece of analytical equipment can separate out a sample's chemical components and determine their purity by comparing the sample to a reference standard. (The cartel's lab may be pretty advanced but normally this purity value is either displayed on a monitor connected to a computer or is calculated later. The rapidly climbing numbers on the digital readout were probably fudged by the production team as a shortcut for viewers.) It might help to think of this separation of chemicals as being similar to the way a prism splits white light into a rainbow spectrum of visible colored light. What starts as white light on one end emerges as seven different colors of light of various wavelengths on the other end. So too does a gas chromatograph split a sample into its various components. The principles at work are different, but the idea is similar.

But if measuring chemical components requires sophisticated machinery, how did Gale manage to measure the purity of Walt's sample without a gas chromatograph on hand? How exactly does this piece of analytical equipment work and what does it measure? And how on Earth could Todd ever hope to measure a sample's purity without the equipment or the know-how? To understand this, I'll have to take this conversation's science level up to the 99th percentile.

Advanced

To understand how the purity of a sample can be measured, we have to first understand just which chemical make-up is preferred in methamphetamine. From there, we can learn the basics of chromatography, how the process separates chemical components in order to better identify them, and how these principles are put to work in the instrument known as the gas chromatograph.

As I mentioned earlier, methamphetamine's chemical formula is $C_{10}H_{15}N$, but it's what's known as a racemic mixture. I talked about chirality a while back in this book, so here's a refresher: Chirality is a description of the geometry of a molecule or ion; a chiral molecule or ion can't be superimposed on its mirror image, usually due to an asymmetric carbon center. The mirror images of chiral molecules/ions are known as enantiomers and are designated as either right-handed or left-handed. Understanding and control of these orientations are crucial to both synthesis and function. A racemic mixture, like methamphetamine, contains equal amounts of right- and left-handed enantiomers, known as levo- and dextro-methamphetamine. (Levo- for levorotatory, or rotating a plane of polarized light to the left/counterclockwise; dextro- for dextrorotatory, rotating it to the right/clockwise. These are often shortened to simply "l" and "d.")

It's not enough just to synthesize methamphetamine. The cooking process needs to be controlled in a manner that produces only the preferred enantiomer. (Walt has a fantastic rant about this in the Season 4 premiere "Box Cutter," in which he rattles off chemistry terminology in order to undermine his possible replacement and solidify his status as a master chemist.) L-methamphetamine, the active ingredient in over-the-counter nasal decongestants, is a stimulant that constricts blood vessels and is active in the peripheral nervous system. However, it has relatively fewer effects on dopamine than d-methamphetamine and does not produce euphoria or have the same potential for addiction. Meth cooks and their clients would therefore prefer a product that is controlled to be all or mostly d-methamphetamine, a process that Heisenberg alone seems to have perfected.

This is all well and good in theory, but it falls to sophisticated analytical techniques to prove out actual "purity" or percentages of chemical components. To that end, chromatography was developed. This technique separates a mixture with the aim of identifying and measuring its chemical make-up. The mixture itself is dissolved in a fluid referred to as the "mobile phase," which carries the sample through a vessel containing another material known as the "stationary phase." Due to differences in the mobile and stationary phases—called "differential partitioning"—and dependent upon the difference in the mixture's chemicals, some of those components will move forward faster than others. "Retention time" is the measure of how long each chemical remains in the stationary phase, which allows each chemical to be identified and analyzed. (Chromatography can also be used to separate a mixture in order to obtain a purified component for later use; this is known as preparative chromatography rather than analytical chromatography.)

To sum the process up in simpler terms, some chemicals in a mixture will be happier staying in the mobile phase (thus moving more quickly through the vessel) and others will prefer to stick with the stationary phase (thus moving more slowly through the vessel). Differences in the components—like molecular weights and boiling points, polarity, and physical size of the molecule—and even differences in the specific traits of the analytical instruments used—mobile phase flow rate, vessel type, and efficiency—will help or hinder the ability to sufficiently separate and analyze the chemicals themselves.[1]

Types of chromatography include column chromatography (in which the stationary phase is housed in an upright tube), planar chromatography (in which the stationary phase is either on a plane or is the plane itself, as in paper chromatography and thin layer chromatography), gas chromatography, and high-performance liquid chromatography, to name but a few. The processes of separating a mixture's chemical components are many and varied and increasingly complex over chromatography's 100-plus-year history, but *Breaking Bad* puts the workload on just one approach: gas chromatography. The basics of using a gas chromatograph are these:

- A sample of material is dissolved in a solvent; you can actually watch the cartel's chemist grind up the sample with a mortar and pestle and dissolve it in some unknown solvent in Season 4, Episode 10: "Salud."
- A small amount of the liquid, on the order of microliters, is injected into the gas chromatograph. (The real-world GOW-MAC Series 580 gas chromatograph appears to be the model used in this scene and is described as the "workhorse" of the lab by the manufacturer.)[2]
- This sample is heated inside the machine to turn it into its gaseous phase, putting the "gas" in "gas chromatography," or the "vapor-phase" in "vapor-phase chromatography."

- The sample gas is then pushed through the machine by a carrier gas—the mobile phase—which is either an inert gas like helium (which does not react chemically under given conditions) or an unreactive gas like nitrogen.
- The sample travels through a long tube, which is coiled to provide more surface area in a relatively small space. This tube, or column, is coated with a thin layer of material—the stationary phase—that the sample can interact with.
- The tube ends in a detector that measures the intensity of the various chemical components that make up the sample against the time it took for each of them to reach the end. This is represented in a series of peaks on a plot known as a chromatogram.

It would be great if, in the real world, this process was actually the way it's portrayed in the show: a quick plug-and-play with instantly calculated results. Unfortunately, real-world tests can take an hour or more just to run through the column. And once you take into account the "heat out" process to remove all of a sample from a column before running samples multiple times for repeatability, the wait time increases even more. Oh, and then there's the process of running a reference sample of the chemical (or chemicals) that the unknown mixture's chromatogram will be compared to ... welcome to lab work in the real world!

And while computers have certainly increased efficiency and decreased the amount of calculations scientists have to struggle through by hand, the multiple chromatograms still have to be analyzed in order to determine the sample's chemical composition and hence, its purity. The first peak often represents the solvent used, while subsequent peaks represent the presence of the preferred chemical(s) and any possible contaminants. After careful analysis, an experienced technician can determine the purity of d,l-methamphetamine in Jesse's sample. (Just imagine the curveball in the chromatogram if Jesse had insisted on using chili powder in his meth cook.) *Breaking Bad*'s shortcut is certainly the preferable version, but unfortunately, it's not a realistic one.

One additional piece of equipment that can help to identify the sample's composition is known as a mass spectrometer. Simply put, this device measures masses of chemical components in a sample. In slightly more complicated terms, the analytical technique ionizes the components of a sample and sorts them by their mass-to-charge ratio. The data that's collected can help to determine the sample's isotopic signature (or ratio of different types of isotopes) and the masses of constituent molecules, and help develop a clear idea of the chemical structures at play. Mass spectroscopy can be used to both determine which compounds of methamphetamine are present in a sample and to determine the ratio of d,l-methamphetamine isomers.[3] Both gas chromatography and mass spectrometry are often used in concert in the lab, and specifically so in drug-testing labs tasked with identifying and quantifying methamphetamine in samples.

So, if Gale Boetticher managed to calculate the purity of Walt's meth sample to 99 percent without using a gas chromatograph, what was he using? Perhaps he had a mass spectrometer squirreled away somewhere in the not-quite-up-and-running Superlab. Assuming he didn't have that equipment ready to go, however, Gale had a couple of other options available to him for determining purity (with one *major caveat* I'll cover later):[4]

Melting point: This is the temperature at which a solid changes to a liquid under atmospheric pressure; the solid/liquid phases exist in equilibrium at this point. The melting point of a given chemical can be used to identify pure compounds in their solid state. This point, which is actually a narrow range of temperatures that begins when the sample starts to melt or soften and ends when it's completely melted, occurs at a relatively high temperature when the sample is pure. Any impurities will both widen the temperature range and lower the melting point itself due to defects in the solid's otherwise stable crystalline structure.

The melting point of a sample can be measured with a variety of devices, such as a metal strip with a temperature gradient known as a Kofler bench or hot-stage microscope, or a Thiele tube that uses hot oil to heat a sample contained in a capillary tube alongside a thermometer. The most common lab device for measuring melting point, however, is known as a Mel-Temp; it uses a magnified eyepiece that allows the viewer to look in on a sample contained in a capillary tube next to a heating element. For the purposes of explaining away the show's narrative, Gale may have used something similar to measure a reference standard of d-methamphetamine—with a melting point of 170°C/338°F—against Walt's sample. Heisenberg's meth must have been close enough to the target melting point to give Gale confidence in his 99 percent purity estimate, but temperatures and times on that small and short of a scale, respectively, would likely fall within a statistical margin of error.[5]

Fractional distillation: Another potential method of analyzing the purity of Heisenberg's meth sample, fractional distillation separates a mixture into its components, known as fractions. To achieve these fractions, the sample is heated in order to get the individual components to vaporize; in other words, they are *distilled* into *fractions*. This method helps to separate components that differ in their boiling points by less than 25°C, like ethanol (78.4°C) and water (100°C), although a simple distillation setup can be used for differences larger than 25°C. Both simple and fractional distillation are common practices in the lab, so Gale would certainly have the knowledge and the resources to carry it out.

Fun Fact: Fractional distillation is also used on an industrial scale in the petroleum industry to separate out components of crude oil into hydrocarbons of different boiling points, which is how we get gasoline, kerosene, and diesel oil, among others, from the same source.

Here's that major caveat I mentioned earlier: These methods of analysis through separation are trying to find the ratio of d,l-methamphetamine in the specific example of *Breaking Bad*. Remember where I mentioned that these two chemicals are enantiomers of each other? That means that they have the same physical properties (melting point, boiling point, etc.) except for their optical activity, in other words, the direction of rotation of a plane of polarized light. This is a complicated way of saying that it's really, really difficult to separate out enantiomers; a melting point or distillation test wouldn't cut it. Gale and the cartel techs would have needed a gas chromatograph-mass spectrometer combo setup at minimum, and for the most accurate measurement possible, the addition of a chiral column. This variant of the stationary phase—the coating of the tube or column—contains one of the enantiomers, which helps to separate out the chiral components based on their affinity for the mobile versus stationary phases.[6]

And now you see why audiences and lab techs would prefer *Breaking Bad*'s simplified version of analytical chemistry! I'll give them a pass for only being about 99.1 percent accurate.

Side RxN #13: Glass Grade

There's a lot of crazy-looking glassware on the set of *Breaking Bad*. Like ... a *lot*! Walt stocks up on the necessaries in the pilot episode of Season 1 and does his fair share of trying to teach Jesse about the proper glassware needed for their meth cooks along the way. Some of it, at least, seems to sink in. In Season 1, Episode 5: "Gray Matter," Jesse attempts to transfer his newfound knowledge to the predictably dense Badger. It's as good an opportunity as any to learn about some fancy glassware ourselves!

Beakers

Griffin beaker—Essentially a glass cup with a pour spout and graduated volume markings, these multipurpose vessels were designed by the nineteenth-century English chemist John Joseph Griffin.

"Volumetric beaker"—Jesse misspoke by calling this bit of glassware a "volumetric beaker," though Badger wouldn't have known the difference. "Volumetric" describes a vessel meant for measuring out a specific volume of liquid at a given temperature. The volume markings on a beaker are only approximations, so it shouldn't be used for accurate measurements. For that, you'd use ...

Flasks

Volumetric flask—Also known as a measuring or graduated flask, this pear-shaped, flat-bottomed vessel with an elongated, narrow neck is calibrated for measuring precise volumes. They're used for preparing standard solutions (a solution made up of a known concentration of elements), for precise dilutions, and as Walt says, "for general mixing and titration."

Round-bottom boiling flask—This self-explanatory vessel is made of heat-resistant borosilicate glass and is used for heating or boiling liquid, distillation, and containing chemical reactions. Due to their round shape, these flasks are often cradled in concave cork rings or clamped to a stand. Walt was *very* excited about the 5,000mL boiling flask in the pilot, but in a later episode, it caught Hank's attention in the high school's chemistry lab storeroom since, as he said, "Meth heads love to brew in this baby." These vessels, like all the other ones on this list, come in a variety of sizes so it's possible to get up to a 20L flask for the lab or even larger for industrial processes.

Erlenmeyer flask—Also known as a conical or titration flask (but let's face it, Erlenmeyer is a much better name), this vessel has a flat bottom, conical body, and cylindrical neck. Named for German scientist Emil Erlenmeyer, who created it in 1860, these flasks are great for titration since they allow for swirling of solutions without spillage. Erlenmeyer flasks are also suitable for boiling liquids and good for recrystallization, though they're not meant for volume measurements. When Hank heads to J. P. Wynne high school, a visit precipitated by finding one of the school's respirators at Walt and Jesse's meth cook site in the desert when trying to track down a missing informant, he also notices that some of these flasks are missing from Walt's chemistry lab storeroom.

Florence flask—Similar in design and purpose to a round-bottom flask, but often with a flat bottom, the Florence flask is used for even heating, boiling, and distillation, and allows easy swirling.

Kjeldahl-style recovery flask—In the mix for the best name ever is this round-bottom flask with a long neck that's useful for trapping splashes from boiling liquids. This flask has its origins in the beer-brewing industry: nineteenth-century Carlsberg brewery chemist Johan Gustav Kjeldahl needed a way of analyzing nitrogen in barley proteins that was simpler than the laborious laboratory methods available at the time. His manner of distillation included this narrow-necked vessel, which would come to bear his name.[7] Walt got particularly excited for an 800mL version of this flask, saying it was "very rare." It's not, really; you can buy a pack of six for less than $200.

Miscellaneous

Vacuum pump—A device that pulls gases out of a sealed container in order to generate a partial vacuum. One was glimpsed in Gale Boetticher's convoluted coffee-maker setup.

Allihn condenser—Also known as a bulb or reflux condenser, this vessel is named after the German chemist, Felix Richard Allihn, and is used for refluxing, or the process of returning vapor condensate to the system that produced it in order to supply energy to a reaction over time. The design, which incorporated a series of bulbs along a long glass tube housed within a water jacket, improved on the performance of the existing condenser of the time.[8] One is also seen in Gale's elaborate coffee-brewing setup.

Autoclave—This device uses heat and pressure to sterilize lab equipment, or in the chemical industry, to cure coatings and vulcanize rubber, among other applications. An autoclave was also included in Gale's coffee-brewing creation for some reason.

XVIII The Lab Maketh the Meth

From Dr. Donna J. Nelson:

Vince Gilligan used different methods to ensure that *Breaking Bad* did not become a cookbook for making dangerous or illegal compounds. He emphatically stated this goal to me multiple times. Methods used included omitting steps in the synthesis, showing multiple syntheses, and seeking assistance from the Drug Enforcement Administration.

Vince sought advice from DEA people for multiple reasons. First, they had firsthand knowledge of illicit syntheses of meth and the equipment used. They would be the best source for the multiple methods he wanted to portray on the show. Second, by taking their advice, the show had their approval for the content displayed. If questions were raised about the suitability of material shown on the show, the DEA could be used as an authority.

The DEA helped the *Breaking Bad* crew with the various illicit meth labs portrayed on the show—the Winnebago lab, the Superlab, the Vamonos Pests lab, and so on. It makes sense that DEA agents and staff would know this, because in real life they dealt with drug makers who created and used such labs in their illicit syntheses.

Sometimes people have asked me whether the *Breaking Bad* crew were actually making meth on the set. The equipment shown were nothing more than empty shells. They were very lightweight and hollow. If one brushed up against a piece of the equipment or tapped it, it made a hollow sound. When walking around in the Superlab, I was always afraid I might bump into something and move it drastically out of position.

We're not going to need pseudoephedrine. We're going to make phenylacetone in a tube furnace, then we're going to use reductive amination to yield methamphetamine. Four pounds.—Walter White, Season 1, Episode 7: "A No-Rough-Stuff-Type Deal"

If you've made it this far, congratulations! You now have the chemistry knowledge needed to follow along with the evolution of Walt and Jesse's meth-making adventures, but also the real-world understanding of what makes this such a dangerous, deadly, and illegal pursuit outside of controlled and certified laboratory settings. Since the previous chapters have acted as a sort of comprehensive "101" section, I'll go ahead and dive right into *Breaking Bad*'s extensive display of meth making, at long last ... right after one more disclaimer and brief digression into the legal history of meth production.

As you probably knew already or have figured out by now, production, distribution, sale, and possession of methamphetamine are illegal in the United States unless prescribed at the federal level. But as Walt mentions in Season 1, Episode 7: "A No-Rough-Stuff-Type Deal," it wasn't always that way. First synthesized in the late nineteenth century, amphetamine and methamphetamine's potent stimulant properties were used to keep soldiers alert and active during World War II. In the 1960s and 70s, amphetamine became popular among athletes, college students, truck drivers, and biker gangs; the latter group soon figured out how to synthesize the more potent, "high-octane" drug methamphetamine from cold medicine after the federal government put amphetamine's precursor chemical ephedrine under controls.

The 1980s saw the spread of the precursor supply from Mexican drug runners to West Coast biker gangs. In 1986, the DEA introduced legislation that intended to monitor manufacturing of chemicals used in illegal drugs. The legislation met with strong resistance from the pharmaceutical industry's lobbying; 1988 saw a compromise between the organizations that created a loophole exempting records of manufacturing of any chemicals—like ephedrine and pseudoephedrine—used for *legal* drugs.

A Cold War of sorts among legislation, law enforcement efforts, and international drug trading exploded in the 1990s in an attempt to cut the flow of precursor while compromising with the pharmaceutical industry's interests. This infighting created more tools for those fighting against illegal meth production as well as loopholes for those perpetuating it. The "Combat Methamphetamine Epidemic of 2005" bill passed as part of the USA PATRIOT Act eventually put restrictions on meth's precursor chemicals, like the then-common over-the-counter decongestants ephedrine and pseudoephedrine. The chaotic and complicated history of meth and the war waged against it set the stage for *Breaking Bad* and allowed the show to play out a condensed version of the real-world conflict's many twists and turns that continue to this day.[1]

Rather than bombard you with a running list of increasingly complicated chemical reactions, I'll be revisiting *Breaking Bad*'s meth-making scenes lab by lab. That's right, we'll take a look at the show's various settings that became characters in their own right. From the RV, to the Superlab, a mobile pest fumigation tent, a desert bunker, and a white supremacist compound, *Breaking Bad*'s meth labs offered up some of the most impressive visuals on the show thanks to very clever writing and an incredibly talented and hard-working production crew. The labs themselves, regardless of how well outfitted they were, provided some of the best scenarios for bringing real-world science into the fictional world of Heisenberg and his associates. So we'll take a look at the labs that made the meth in a moment, but first we'll check out the meth itself and its drug-free substitutes used for the show.

Inside *Breaking Bad*: As you're about to discover, meth can be made in myriad ways. The show's stand-ins for the drug likewise have multiple sources. In the pilot episode, Gilligan said that the meth was "a Japanese packing material" according to his prop master at the time; this is the same stuff they use for stunt sequences in which a clear rubber material stands in for broken glass.[2]

From that point on, the meth was actually just sugar with some food coloring—in other words, rock candy. (Despite this, the actors were still not allowed to be shown inhaling any of the smoke ... not that they would have enjoyed doing so anyway.[3]) It was supplied by Debbie Ball, "The Candy Lady" of Albuquerque, also known as "The Bad Candy Lady," who was initially approached during the show's first season by *Breaking Bad*'s prop department. Ball supplied both the initial white crystal rock candy (cotton candy-flavored) and the Blue Sky version, which she now sells on her website (drug-free, of course[4]).

Gilligan admits to having a vague idea of coloring the meth light blue, like the sky, towards the end of Season 1 so that the locals might call it "Sky"; the fact that Walt's wife's name is Skyler would have tied into this theme in a troubling manner, but ultimately nothing came of the connotation.[5]

Cap'n Cook Gets Busted

Oh, Jesse Pinkman, how far you've come in five seasons of *Breaking Bad*. Before Walter White ever gets the idea to synthesize his own meth (and well before his alter ego Heisenberg gets designs on building a drug empire), his former student Jesse Pinkman was a small-time meth user, manufacturer, and dealer operating under the name Cap'n Cook. (Don't hold your breath for a *Breaking Bad* prequel series for this character.) Cap'n Cook's claim to not much fame was adding a bit of chili powder to his cook in order to make his "Chili P" product stand out to users. While Jesse probably would have been happier to continue as Cap'n Cook, his customers much preferred the new product brought to market by the mysterious genius known only as Heisenberg.

But while Walt and Jesse's RV lab was technically the first one shown on *Breaking Bad*, chronologically we have to go back in time a little bit further. As fate (and the writers' room) had it, Walt's ride-along with his DEA agent brother-in-law Hank to bust Cap'n Cook's meth lab actually set him on the path to becoming Heisenberg and partnering up with Jesse Pinkman, way back in the series' pilot. (In what could seem like a throwaway line, Walt corrects Hank that it's phosphine gas and not mustard gas that can be produced while synthesizing methamphetamine. This short bit of dialogue served to establish Walt's understanding of chemistry early on and foreshadowed a gruesome bit of action to come, which I talked about back in chapter 4.)

It suffices to say that this meth lab—set up in a neighborhood home, no less—is a mess. Dirty bottles and equipment line the countertops, various chemicals and solvents—including toluene, iodine, and nail polish remover—lay out in the open, and partially used packs of cold medicine are scattered everywhere. Other paraphernalia includes road flares, funnels and coffee filters, matchbooks, and jugs full of clear, yellowish, and red liquids. This all might look pretty random, as if the production team had simply taken whatever items they could find from their kitchens and garages and used them to decorate the set. Not so. The production team certainly did their homework. In fact, the use of common, everyday items in the synthesis of illicit methamphetamine is what makes the illegal activity relatively easy to perform but difficult to combat.

After the cooking culprit (and Jesse's original meth-making partner) Emilio Koyama is taken into custody by the DEA, Walt presumably gets a better look at the cook site off camera, but audiences are left with the fleeting images of the messy meth lab from the DEA's raid. The brief look is more than enough to piece together the fact that Emilio's crude approach to methamphetamine synthesis is based on the method of nineteenth-century Japanese chemist Nagai Nagayoshi and twentieth-century Japanese chemist and pharmacologist Akira Ogata. A fantastic run-down of the show's meth-making chemistry was laid out by chemist Jason Wallach in a 2013 *Vice* article, which I'll be referencing quite a bit in this chapter.[6]

After the discovery and synthesis of amphetamine by Romanian chemist Lazar Edeleanu in 1887, Nagai was able to synthesize methamphetamine from ephedrine in 1893. This achievement grew out of his study and chemical analysis of traditional herbal medicine while at Tokyo Imperial University where he was able to isolate ephedrine from a plant in the *Ephedra* genus.[7] A stimulant, ephedrine has been used to control asthma, blood pressure, and body weight, but in our discussion, it acts as a methamphetamine precursor. In 1919, Akira Ogata first synthesized methamphetamine hydrochloride, a.k.a. "crystal meth," by reducing ephedrine using red phosphorus and iodine, a method that is still used today in illegal drug manufacturing.

Remember how l-methamphetamine and d-methamphetamine are enantiomers of each other? So, too, are ephedrine and pseudoephedrine ($C_{10}H_{15}NO$); technically each designation is actually a *pair* of enantiomers since the molecule has two chiral carbons compared to methamphetamine's one. The only thing separating these chemicals from methamphetamine is one hydroxyl group (-OH), which meth cooks and chemists would like to remove.

Now's as good a time as any to make a distinction between a "cook" and a "synthesis" as well as the difference between meth making as chemistry and as "art," which Pinkman claims it is in the pilot. Consider a "cook" like microwaving a burrito: You can follow the written instructions without any formal education or experience, as long as you have the right equipment, and just hope and pray that it doesn't blow up in your face. Now consider "synthesis" like a world-class chef striving to create the perfect meal: Years of training and experience allow the chef to get the necessaries right every time along with the confidence to add her own signature style and respond to any alterations in the ingredients, equipment, or the process itself. Both methods result in something edible, but of vastly different qualities. (I'll continue to use them interchangeably throughout, however, since it's just so much fun to say "cook.")

And as for the "art" of meth making, as in food preparation, that's in the eye of the beholder or the preference of the end user. Some folks are perfectly happy with fast-food fare while others will only settle for a $100 deconstructed burger topped with a hickory-smoked foam on a gluten-free bun, so whether it's Chili P or Blue Sky, if your chemistry skills are solid, there's more latitude for artistic experimentation.

Back to the synthesis. The original early twentieth-century methods may have used ephedrine as the precursor that was reduced to methamphetamine, but the show opts to use pseudoephedrine instead; the reduction process, using hydroiodic acid (HI) and red phosphorus, is the same. These ingredients may sound somewhat exotic, but remember those common items in the lab? They're the easily obtained sources of the necessary chemicals:

- Pseudoephedrine can be extracted from over-the-counter cold medicine (explaining the coffee filters and jars of extracted liquid; the red liquid comes from filtering the medicine's red wax coating).
- Red phosphorus, used to reduce iodine to hydroiodic acid (I_2 -> HI), can be collected from the matchbooks and road flares. As Heisenberg himself tells a wannabe meth cook in Season 2, Episode 10: "Over," "Red phosphorus is found in the striker strips, not the matches themselves."
- Iodine crystals, which are perfectly legal in and of themselves, are regulated by the federal government because of their potential use in illegal methamphetamine production, but can also be extracted from disinfectants or antiseptics known as iodine tinctures.[8]

These ingredients, when combined, heated, filtered, adjusted for a specific pH, extracted, and crystallized in a somewhat lengthy process eventually produce methamphetamine. Simply, this is done by kicking off that pesky -OH group and replacing it with hydrogen, turning cold medicine into a powerfully addictive drug. (It's amazing what one little tweak of chemistry can do.) But more accurately, the introduction of HI produces a reaction known as a nucleophilic substitution. In this type of atomic exchange, the introduced chemical's electron pair "attacks" or selectively bonds with a relatively positively charged atom, displacing it as a "leaving group." In this case, the iodide atom [I^-] of HI is the nucleophile, or attacker; it bonds with the carbon attached to the hydroxyl group, causing the OH^- molecule to depart and soon form H_2O with the available H^+/H_3O^+ in solution.

To kick that newly bonded iodide atom off and replace it with the desired hydrogen requires a second step. The C-I bond is the weakest of the carbon-halogen bonds due in part to iodide's relatively large atomic radius, low electronegativity, and low bond-dissociation energy. It's the best leaving group of the halides, even though it's also the "attacker" in the reaction just described. The C-I bond is so weak that organoiodine compounds often take on a yellowish appearance due to the formation of I_2, an impurity formed by available iodide atoms in solution.

However, this formation of I_2 works to the advantage of meth synthesizers. Iodine prefers to be in its elemental state (I_2), in which two iodide atoms share an electron pair in a strong covalent bond, rather than the relatively weak C-I or HI pairings. This characteristic not only helps to remove iodide from the intermediate molecule in order to form methamphetamine, it also allows for the regeneration of more HI from the newly formed I_2, which is used to propagate the first part of the reaction in other pseudoephedrine molecules. That's where the red phosphorus comes in.

The phosphorus (P_4) reacts with the I_2 in the presence of water (some of which is produced from the initial reaction when the hydroxyl group is kicked out) to make PI_3, or phosphorus triiodide. This chemical reacts rather vigorously in water to produce phosphorous acid (H_3PO_3) and the sought-after HI. Then, the process can start all over again within the reaction vessel. There's a lot going on at the atomic scale in this (and every) reaction; understanding the complicated process means the difference between being a cook or a chemist.

As far as theoretical yields go, this method can produce a final product weighing up to 92 percent of the weight of the precursor, though illicit labs tend to range from 50–75 percent yield due in part to the lack of necessary skills and equipment. Walter White has both, so let's see what happens when he takes his first crack at making meth in the fan-favorite RV, "The Crystal Ship."

The RV

Let's take a moment to appreciate the RV, shall we? As a mobile meth lab, it's an ingenious bit of writing. And as a functioning and recurring set piece, it's a clever bit of production work, as well. The interior of the actual RV model, a 1986 Fleetwood Bounder, was too small to accommodate the cast and crew, so the production team made a set to stand in for it.[9] The RV was the setting for a number of dramatic events throughout the series as well as the literal driving force behind Walt and Jesse's burgeoning meth empire. It was their base of operations all the way up to the halfway point of Season 3 when it was destroyed in order to eliminate incriminating evidence. (RIP, Crystal Ship, you will be missed.) In memoriam, let's relive some of the RV's finer meth-making moments.

In the pilot, Emilio Koyama isn't the only chemist to employ the Nagai method described earlier. Walter White himself uses this process in his first cook session with Jesse in their newly christened mobile meth lab. Audiences get to enjoy the brilliantly edited meth-making montage in this episode, thanks to the work of editor Lynne Willingham and then-assistant editor Kelley Dixon. (Dixon also hosted AMC's *Breaking Bad Insider Podcast*, the fantastic behind-the-scenes show from which many production insights are shared in this book.) In the montage, the lab itself is clearly more organized and much cleaner than Emilio's countertop operation, though it's still outfitted with pilfered high school chemistry lab equipment and less-than-stellar safety precautions. It's no Superlab, but it's a start.

In this montage, Walt and Jesse are shown grinding up pills with mortars and pestles (a classic lab maneuver), dissolving them in alcohol and extracting the pseudoephedrine through filtering, followed by the addition of red phosphorus and iodine crystals. After the mixture is heated to boiling (and boiling over, thanks to Jesse), Walt filters the resulting mixture into a beaker. You can clearly see him testing the solution's pH to adjust it accordingly before proceeding with extraction via an organic solvent, which he then removes with the use of a syringe; this could also have been evaporated off. (The *Vice* article cites the use of a separatory funnel—which is designed for separation of liquids of different densities during extraction—as an easier method favored by chemists.) Hydrogen chloride gas is then bubbled into the leftover liquid that contains methamphetamine and solvent, a process that precipitates the desired d-methamphetamine as an HCl salt, seen as a white paste. What's not shown in the montage is the process of filtering and drying these crystals, though the crystal-clear, "glass-grade" final product is given the spotlight.

One of the problems with this method of meth making is obtaining the supply of precursors. By Season 1, Episode 7: "A No-Rough-Stuff-Type Deal," the frustrating and lengthy process of "smurfing" pseudoephedrine—employing people to buy small amounts of over-the-counter cold medicine from a variety of stores across a large area—becomes untenable. Walt's in the empire business, after all. To get around this snafu, Walt opts to switch his cook to an alternate method: reductive amination using phenylacetone and methylamine, otherwise known as the P2P cook. To quote Jesse Pinkman, "Yeah, science!"

A P2P cook, taking its name from the chemical phenylacetone ($C_9H_{10}O$, a.k.a. phenyl-2-propanone, or P2P), is a more complicated process with less room for error and a lower overall yield than a pseudoephedrine cook, but the precursors are relatively easier to come by in sufficient quantities. Superfans will know that procuring methylamine becomes a major plot point throughout the seasons, a difficulty introduced in this very episode. (Head on back to chapter 9 to revisit the thermite lockpick if you need a refresher.) But the precursor isn't the only thing Walt will need to change up his approach. The shopping list he gives to Jesse includes the following:

Autotransformer—Used to regulate voltage from a power supply, stepping it up or down to the required power level for a piece of equipment, such as a tube furnace.

Thirty-five-millimeter tube furnaces (two) or 70-millimeter tube furnace (one)—An electric heater, usually a cylindrical cavity surrounded by heating coils within an insulator, used to synthesize and purify compounds.

Anhydrous methylamine (6L)—Precursor chemical (CH_3NH_2)

Thorium nitrate (40g)—A radioactive chemical compound, $Th(NO_3)_4$, used to produce the catalyst, thorium dioxide (ThO_2).

Hydrogen (electrolytically produced)—Electrolysis is the process of passing an electric current through water to split it into oxygen and hydrogen gas. (It's the reverse of the process I talked about in chapter 5 regarding Walt's DIY battery.) This removes or at least reduces chances of contamination.

Now, I know that the RV technically wasn't the site of the first P2P cook—it had broken down, as it often did, forcing Walt and Jesse to temporarily cook in Jesse's own basement in the Season 1 finale—but it becomes the main cook site for Blue Sky meth for the next season and a half, so I'll cut it some slack here.

After boosting the barrel of methylamine—which should keep Walt and Jesse cooking "for the foreseeable future"—the drug-making duo set to work on their new method. The setup looks much more stripped-down and bare-bones in the season finale than the pseudoephedrine cook scenes glimpsed earlier in the season since the equipment and glassware just sit atop a pair of folding tables. We're still a long ways off from the Superlab.

However, the chemistry is still sound. A reductive amination reaction involves the following steps:

- A ketone or aldehyde—simple compounds with a carbonyl group, that is, a carbon-oxygen double bond, RC(=O)R′—is reacted with an amine, a compound that contains a basic nitrogen atom.
- The two molecules combine into one larger one, along with the loss of a smaller molecule. This is known as a condensation reaction.
- The newly formed molecule is known as an imine—a compound containing a carbon-nitrogen double bond (C=N)—or a Schiff base, if that nitrogen atom is not bonded to a hydrogen atom.
- This is then reduced (i.e., it gains hydrogen) to form the desired amine.

In other words, reductive amination is an impressive way of saying, "Combine two molecules to make an imine. Turn it into an amine." This is a powerful and versatile process in the chemist's tool kit, especially since it's more controllable than alkylation—the transferring of alkyl groups (like $-CH_3$ or $-CH_2CH_3$, etc.), which can lead to a mixture of undesirable products—and both steps can be done in the same reaction flask.

When it comes to *Breaking Bad*, reductive amination is the path to Blue Sky, Heisenberg's signature crystal meth. It makes its first, unforgettable appearance in the Season 1 finale. Following along with the basic steps I've outlined, the phenylacetone (P2P) is the ketone, which is combined with the amine precursor, methylamine. The resulting imine is then reduced through the addition of hydrogen to produce a racemic (50/50) mixture of d,l-methamphetamine. Simple, right?

Well, not quite. First of all, there are a number of methods that can be used to reduce the imine. Options include sodium cyanoborohydride ($NaBH_3CN$); using hydrogen gas (H_2) and a precious metal catalyst like platinum dioxide (PtO_2), known as Adams' catalyst; Raney nickel, a derivative of a nickel-aluminum alloy; or an amalgam of aluminum and mercury. It's this latter reductive agent that Heisenberg eventually opts for, as referenced throughout the show.

In Season 3, Episode 4: "Green Light," Jesse displays the surprising fact that he's learned something from all the cooking sessions with Walt. When questioned by a smug Heisenberg as to what Jesse used for the reduction step in his own off-brand meth making and accused of using platinum dioxide, Jesse says that he in fact used the aluminum amalgam because "the dioxide's too hard to keep wet." This comment might have been about how insoluble platinum dioxide is in water, but it's more likely that it referred to keeping the platinum catalyst itself, known as platinum black, away from the air or other oxidizing agents, since it is water-reactive and also pyrophoric, meaning it will ignite spontaneously in air below 55°C/130°F. Not a particularly fun material to be working with on a regular basis. There is even evidence that Walt attempted to use the Adams' catalyst as a reductive agent in his first P2P cook since hydrogen gas, which is used with this catalyst, was on his shopping list, but the aluminum amalgam method won out in the end.

There are some other things complicating the process of the P2P cook. Phenylacetone, the chemical that gives the process its name, was designated as a Schedule II Controlled Substance by the DEA back in 1980. That's not stopping Heisenberg since P2P can also be synthesized a number of ways. And thanks to Walt's shopping list and some dialogue throughout the series, we know that he's synthesizing it from phenylacetic acid ($C_8H_8O_2$), or PAA. (That, too, is on the DEA and U.S. Attorney General's watch lists of controlled chemicals.)

The dead giveaways on the list that point to PAA-to-P2P synthesis are the requests for thorium nitrate and a tube furnace. As I mentioned earlier, thorium nitrate is used to generate a radioactive metal oxide catalyst known as thorium dioxide (ThO_2). This catalyst bed, heated by the tube furnace, receives the vaporized forms of PAA and plain old acetic acid ($C_2H_4O_2$, the main ingredient in vinegar besides water). These acids, known as carboxylic acids (for their carboxyl group, -COOH), go through a rather complicated chemical reaction to ultimately form the desired ketone, P2P. (There are also undesirable impurities in this reaction, including symmetric ketones like acetone and dibenzylketone, among others, as well as the side products of carbon dioxide and water.) The resulting product, a brownish oil, is then separated from the aqueous layer before the P2P itself is collected and purified through vacuum distillation.

Walt and Jesse may have their P2P synthesis and reduction method locked down, but the availability of the *other* precursor, methylamine, remains a thorn in their sides the entire series. The writers simply could have let Heisenberg make his own methylamine as well, but then we would have been denied the drama of Walt and Jesse's lock-burning warehouse robbery; high-stress business dealings and negotiations with suppliers, like the Mexican cartel, Gustavo Fring, and Lydia Rodarte-Quayle; and, it goes without saying, the incredible and intense train heist.

Methylamine can be made industrially by reacting ammonia with methanol using an amorphous silica-alumina catalyst,[10] and it has also been made in the lab by hydrolysis of the extremely toxic and hazardous chemical methyl isocyanate. (Nineteenth-century French chemist Charles Adolphe Wurtz discovered the actions of isocyanates as they relate to urethane chemistry, specifically preparing methylamine by this process in 1849.[11]) Another process, known as the Hofmann rearrangement, after the excellently named nineteenth-century German chemist August Wilhelm von Hofmann, synthesizes methylamine by reacting acetamide in a basic solution with bromine, after proceeding through a number of intermediate chemicals. So while it was possible for Walt to make all of the precursor he ever needed, it probably wasn't cost effective or timely to do so. And it certainly wasn't as dramatic as the *Breaking Bad* we know and love.

So even if your head's spinning from all of this advanced chemistry talk, hopefully you can now appreciate the meth-making montages a little more. The one that takes place in the Season 1 finale is, unfortunately, cut short by an impromptu open house event led by Jesse's realtor, but there's still plenty of basement meth making on display. The tube furnace is in plain sight on top of the folding table—which should be used to synthesize the

P2P, not the final methamphetamine product, remember—along with a condenser setup that presumably helps to control the very vigorous reaction and/or collects the product vapors that form after cooking the mixture at "425 degrees, running the process for a minimum of 2 hours to make enough for 4.5 pounds," according to Walt. The most eye-catching thing in the lab, however? The sky-blue meth.

"Blue Sky" certainly grabs the attention of Tuco Salamanca, the first drug dealer to sample the potent product. While Heisenberg's meth soon rockets to the top of the charts as far as meth aficionados are concerned, the blue color is likely more for dramatic flair than anything based in real-world chemistry. One-hundred percent pure methamphetamine in its HCl salt form is a colorless-to-white crystalline solid, with street-produced meth existing in a range of colors, namely colorless, white, and yellow. Any color present in crystal meth, whether it's crystal clear blue or an opaque and crumbly brown, is due to impurities that accumulate during the reaction.

Still, in the *Breaking Bad* narrative, Walt's meth is the top of the tops, color-imparting impurity or not. Perhaps the color isn't just a result of the change in Heisenberg's synthesis method, but an extra tweak of chemistry that makes what should be a racemic mixture enantiomerically pure. Walt does have the devil's luck, after all. But luck will only get him so far. To start building an empire, he's going to need a Superlab ...

The Superlab

If you made it through that chemistry-heavy section, the good news is that the process stays pretty much the same from here on out; the cook sessions are simply colored by the much-improved lab setup that Walt and Jesse find themselves in. The bad news is that this is your final opportunity to bid a fond farewell to the Crystal Ship. The last we see of the RV, chronologically, is in Season 3, Episode 6: "Sunset," in which the Bounder is destroyed beyond recognition in order to prevent Hank and the DEA from gathering some very, very incriminating evidence. It's referenced a couple more times in nostalgic moments between Walt and Jesse, and it's glimpsed one last time in a flashback seen in Season 5, Episode 14: "Ozymandias." In that hour, audiences revisit Walt and Jesse's first cook in which the teacher is trying to educate his former student, saying: "The reaction has begun. If we had a freezer, the condensed liquid would cool more quickly because, of course, this is an exothermic reaction, which means 'giving off heat' and our desired product is a crystal ..." Jesse may not have appreciated this moment at the time, but fans certainly did.

However, fans also appreciated that Walt and Jesse (eventually) got to move up in the world of meth making by taking up residence in Gus Fring's clandestine facility dubbed the Superlab. This was a gorgeous setting despite all the ugliness that would take place there. Vince Gilligan's original thematic idea for the Superlab was that it should feel like a bunker. It fell to production designer Mark Freeborn to design the lab, while the actual building of the two-story, practical set was carried out by construction coordinator William "W" Gilpen and their crews.[12]

Built to look like it's made out of board-form concrete, the set actually used a thin veneer of plaster to mimic the bunker-like aesthetic. Parts of the walls themselves could be removed to accommodate cameras and crane placement. The steel staircase and catwalk were built to code[13] and the lab itself was outfitted so that you could literally cook meth on the set, per Gilligan.[14] Not that they ever did, of course.

From a writing standpoint, the Superlab is the next step in the clever evolution of the show's meth-making facilities that allow Walt and Jesse to carry out their cooks while also offering protection from discovery. The cover story in this case is provided by the industrial laundry above the lab since it receives regular chemical deliveries and vents "clean, odorless steam" thanks to an overengineered filtration system. Ingenious. Although this fantastic set was eventually dismantled, the production team had to build a duplicate in order to show the Superlab in its destroyed state when it's being combed over by the DEA in the Season 5 premiere, "Live Free or Die."[15]

In its prime, the Superlab was a work of science-loving art. Though the Superlab itself is first revealed to the audience in Season 3, Episode 5: "Más," chronologically, we get to glimpse Gale unpacking newly delivered lab equipment and gushing to Gus over the lab's quality in the Season 4 premiere, "Box Cutter." As previously mentioned, the expensive lab equipment on display here is noted as being at home in the world's most successful pharmaceutical companies, but without the world's best meth chemist, all the equipment in the world can only get you so far. (It's the difference between Gale's 96 percent purity and Heisenberg's 99.1 percent, after all.)

Efficiency is the name of the game for Heisenberg, which is why his attention to detail and knowledge of chemistry are particularly attractive to the more business-oriented drug lords in *Breaking Bad*. The average meth user may not notice much of a difference between 96 percent and 99.1 percent, but there's a good chance that the end user won't get either level of purity from a street-level buy. One reason that Walt's nearly perfect meth is so desirable is that it is somehow enantiomerically pure, providing the preferred d-methamphetamine without its less desirable chemical counterpart.

Another reason why Heisenberg's synthesis is efficient is that, once you begin to account for economies of scale, Walt's attention to detail is likely saving Gus millions of dollars both on the supply side, since the precursor chemicals are being efficiently used, and on the demand side, since he can charge a premium for Blue Sky. Gus's extensive distribution network also benefits from selling the 99.1 percent meth since there is more opportunity for that product to be diluted—"cut" or "stepped on"—in order to sell a greater volume of meth that's lower in purity for a lower price. Just like gasoline is available in Regular, Plus, and Premium, the higher-end choices may give you more bang, but it'll also cost you more bucks.

But that's enough economics: let's get back to the science! In the very same episode that introduces the Superlab, Walt confirms his use of some of the chemistry we've mentioned. Like a kid unwrapping birthday presents, Walt is in a state of pure bliss as Gus allows him to run free in the Superlab. The first package he reveals is "thorium oxide for a catalyst bed," which is used to synthesize phenylacetone (P2P), as we discussed earlier. The massive reaction vessels—1,200L—will allow Heisenberg to up his production volume quite a bit, much to Gus's pleasure since he needs on the order of 200 pounds per week to make the Superlab a viable meth-making facility. However, Walt will still need an assistant for the complex and much-expanded meth-production process.

Enter Gale Boetticher. Arriving in Season 3, Episode 6: "Sunset," this capable assistant impresses Walt with his knowledge of chemistry—and makes a damn fine cup of coffee to boot. (See chapter 7. As an aside, Gale's lab etiquette and attention to detail may be off the charts, but it's never a good idea to have beverages in the same space as toxic chemicals. Or to wear open-toed shoes in the lab ...) The sandal-wearing assistant with a bachelor's degree from the University of New Mexico and an organic chemistry master's degree from the University of Colorado—with a specialization in x-ray crystallography, of course—is a big step up from Jesse, at least in the arena of scientific knowledge. And you know what that means: Time for a meth-making montage!

In their first cook together in the Superlab, after a cup of coffee, Walt and Gale don their bright yellow hazmat suits and respirators as they get to work. They start off by grinding a dark red powder in mortars and pestles (a running gag at this point considering that they're working with volumes far too great to be grinding things by hand, but at least they're doing it in unison like synchronized swimmers), followed by dumping aluminum foil into a reaction vessel that probably contains mercury (II) chloride in water to form the reducing agent, aluminum amalgam. The scenes of Walt adding drops of a red liquid to various test tubes (probably to keep track of pH, as indicated by the values on the whiteboard on the wall behind the pair's ongoing chess game) and Gale using a thin capillary tube to capture a sample, likely for thin-layer chromatography or TLC (see chapter 17), are interesting additions to the montage since they reveal the pair's quality control steps throughout the process. Clearly, this is a serious synthesis, not some basement cook session.

Another clue that Walt is still rocking the P2P cook at this point is Gale's question about the phenylacetic acid solution, asked during a celebratory glass of wine after the trays of Blue Sky have been placed in their drying racks. He wants clarification on the procedural step of "150 drops per minute for the first 10 minutes, and then 90 for the remainder." Walt's answer is that, "by tapering the phenyl you get an oilier aqueous layer," which gives "better benzene extraction," as Gale understands it, finishing Walt's sentence. (Walt, however, prefers ether.)

Reading between the lines here, they're talking about the synthesis of water-insoluble/sparingly soluble phenylacetone (P2P) from the relatively more water-soluble phenylacetic acid (PAA) and acetic acid (AA). The reaction vessel, containing both compounds, will see a separation into non-mixing aqueous and organic layers, with PAA/AA in the former and P2P forming and migrating into the latter, which can be separated and collected relatively easily. The issue is that this layer will also contain some of the partially soluble PAA that remains unreacted, which is undesirable in the final product. By slowing the rate of addition of the PAA later on in the reaction, Walt is trying to control the rate that the reactants are being converted into the product, avoid side reactions, and maximize the amount of PAA that gets converted into P2P. If, somehow, the mixture's aqueous layer could stay "oilier" longer, any remaining unreacted PAA with its non-polar benzene ring might be more likely to stay in this layer instead of the organic layer containing the desired product. Keeping the PAA surrounded by acetic acid ensures that only these two molecules will react with each other, which is the desirable outcome.

As for the comment that this change gives "better benzene extraction," they're talking about a follow-up step after separating out the aqueous layer. Benzene can be used as a nonpolar solvent in a number of alternate synthesis methods—which would explain why Walt mentions his preference for ether, a very common nonpolar solvent—though toluene is preferable since its solvent properties are similar, but it's less toxic to handle. It's likely that Walt is aiming to increase the yield through a benzene extraction, though it's possible that he is actually employing a synthesis that's completely novel, or perhaps the *Breaking Bad* writing team opted to pick and choose from a variety of methods in order to throw would-be cooks in the real world off the scent. Either way, it's always a pleasure hearing this talented cast deliver scientific terminology with ease.

Walt and Gale's chemistry skills easily are the most advanced on the show, but their *relationship* chemistry leaves a lot to be desired. They're like classical music versus jazz, or perhaps oil versus water is more appropriate. It's not long—the very next episode, in fact—before Gale starts to get on Heisenberg's nerves. That's a very dangerous position to be in. So in Season 3, Episode 7: "One Minute," despite Gale going to great lengths to provide everything Walt needs (the trays are clean and dry, the solvent's been filtered) and to rectify his own shortcomings (he's wearing closed-toed shoes this time since they're "more professional"—and safer, Gale), Heisenberg loses it over a temperature error.

While Walt is correct that the difference between 75°C and 85°C is quite the gulf when it comes to their procedure and quality control, especially when fifty gallons of product go to waste due to the deviation, he's clearly got an ulterior motive here. Heisenberg needs Gale out of his way and out of the lab, first, because Gale is competent enough to master the cook, which would make Walt irrelevant and expendable; second, because Heisenberg likes being in control; and third, because Jesse Pinkman is a loose end who needs to be brought back under Walt's wing. This all happens in the very next episode, "I See You," as Gale is ousted and Jesse takes his place. It's a very awkward and difficult break-up scene, but worse things are to come for Gale ...

Jesse and Walt are back at it in the Superlab for a time—where Jesse regularly lifts supposedly 200-pound containers of meth with relative ease, by the way—along with more cooking sequences and montages. But it isn't long before circumstances force Jesse to go into hiding, meaning Walt and Gale team up once more. This time, however, Gus's associate Victor watches their every step and Gale seems more nervous and inquisitive than ever, asking questions he should well know the answer to, like if there's a trick to purging the catalyst bed. Unfortunately for Gale, he finds himself caught between Heisenberg and Gus, an untenable position that ends with his death in the Season 3 finale: "Full Measure."

But there's still one more gruesome scene left to play out in the Superlab. Gale may be gone, but Walt and Jesse are on the hook for his murder, and as far as Gus is concerned, for maintaining the cook schedule. Wily Walt thinks he's played a trump card in killing Gale, but Victor has been paying close attention, close enough to replicate Walt's cook step for step. With only one trick left to play, Walt unloads a barrage of chemistry conundrums aimed at confounding the would-be cook and forcing Gus to see the error of his ways before it's too late.

Here's Walt's epic rant (Season 3, Episode 13: "Full Measure") in full:

> Please, tell me, catalytic hydrogenation: Is it protic or aprotic? Because I forget. And if our reduction is not stereospecific, then how can our product be enantiomerically pure? I mean, it's 1-phenyl-1-hydroxy-2-methylaminopropane containing, of course, chiral centers at carbons number one and two on the propane chain, then, reduction to methamphetamine eliminates ... which chiral center is it again? Because I forgot. C'mon, help me out, professor! ... What happens when you get a bad barrel of precursor? How would you even know it? And what happens in the summer when the humidity rises and your product goes cloudy? How would you guard against that?

Let's pretend we're Victor and play along with Walt's trivia challenge, answering them one by one. First up, "catalytic hydrogenation": this is a fancy way of saying that you want to add hydrogen to something in the presence of a catalyst, an agent that speeds up the reaction. Walt is referencing his reductive amination chemistry, specifically the final step where hydrogen is introduced into the imine to reduce it to the final d,l-methamphetamine product. Since "protic" solvents are "proton-donating" and have a hydrogen atom bound to either oxygen or nitrogen ("aprotic" is not proton-donating and the solvent is absent such a bond) and since the reaction adds a proton in the form of hydrogen, the answer is: protic.

Walt's next question deals with stereospecificity and the enantiomeric purity of his process. Again, this is a fancy way of saying that the procedure needs to ensure the correct products are formed at the reaction's end. (Stereoisomers are simply molecules that have the same composition and arrangement of atoms but differ in their three-dimensional structure. Enantiomers are stereoisomers that are also non-superimposable mirror images of each other.) A stereospecific reaction *specifies* the form of a given product for a given reactant; if the reactant is stereoisomerically pure, then a stereospecific reaction will produce 100 percent of that specific stereoisomer. A stereo*selective* reaction, on the other hand, allows for multiple products to be formed, but favors one particular stereoisomer depending on a number of factors. Ideally, Walt's reaction would start with the pure stereoisomer of P2P to undergo a stereospecific reaction that would ensure the proper products are produced.

The issue is that P2P isn't chiral (it's technically prochiral, meaning it can change from achiral to chiral in a single step) and neither is the resulting imine from the reaction with methylamine. Therefore, this formation reaction and the following reduction reaction are not stereospecific. However, such mechanisms do exist. A subset of highly controlled chemical reactions for the synthesis of methamphetamine came into vogue post-1985, though it's unclear if that's the same science playing out behind the scenes of *Breaking Bad*'s cook sequences. Approaches include the use of exotic-sounding chemicals like chiral organometallic ligands with transition metals, chiral oxy-alkylation of 1-phenylpropanal with nitrobenzene, and the use of acetic anhydrides in ester production, followed by catalytic reduction.[16] That's far beyond the scope of this book (or this particular writer), but the science exists.

But the more important words in Walt's rant are "enantiomerically pure." Remember that the l-methamphetamine enantiomer does no one in the meth business any good since it's basically a decongestant. It's the d-methamphetamine form Heisenberg is after. Curiously, his earlier pseudoephedrine reduction method would actually produce more of this preferred stereoisomer since the chirality of the precursor would be preserved in what's called an asymmetric or enantioselective synthesis. Instead, the later reductive amination cook produces the racemic (50/50) mixture of both stereoisomers. It's possible that Walt has found some way of selecting against the undesirable enantiomer during the reaction, or that he's simply discarding the unwanted 50 percent of his final mixture; either way, that's not great for his yield. His enantiomeric purity, however, could still be 99.1 percent.

On to everyone's favorite chemistry subject: nomenclature! Walt's mouthful (1-phenyl-1-hydroxy-2-methylaminopropane) is one way to name ephedrine/pseudoephedrine, though the IUPAC (International Union of Pure and Applied Chemistry) naming convention is: (1R,2S)-2-(methylamino)-1-phenylpropan-1-ol. (I'll stick with "pseudo," thanks.)

Remember: to get to amphetamine, we have to reduce pseudoephedrine and get rid of that pesky [–OH] group. Well, that's located at carbon number one, which loses chirality when the hydroxyl is removed and replaced with a hydrogen atom. (Why Walt is talking about his old pseudo cook when Victor is in the process of carrying out reductive amination is beyond me, but I'm guessing it's an attempt to fluster the would-be replacement.)

At this point, Victor still hasn't missed a step in the cook, regardless of the fact that he doesn't know a lick of chemistry beyond the steps listed in the procedure. This is where Walt changes up his approach by throwing theoretical wrenches into the works. When something goes wrong with your ingredients or the recipe doesn't address a curveball, what do you do? Well, for a bad batch of precursor—a quality control issue, really, which is why chemists buy their supplies from reputable and accountable distributors and perform quality checks—Victor could always run a sample through a gas chromatograph or other analytical piece of equipment to compare it against a standard and records of previous in-house measurements. As for humidity, proper climate control in a facility like the Superlab should negate any environmental factors. However, excessive humidity can cause anhydrous chemicals and desiccants used in the reaction to pull more moisture from the air (and less from the reaction), as well as lengthen the drying process.

If you successfully made it through Walt's chemistry gauntlet, you deserve a pat on the back. This is Walt at his most Heisenberg-ian. Unfortunately, Victor never gets a chance to respond in a sufficient manner since Gus puts an end to him. Walt and Jesse may have lived to see another round of meth making, but Gus's henchmen Tyrus and Mike start to watch over the pair in Season 4, Episode 2: "Thirty-Eight Snub." More drama unfolds over the course of Season 4, though the science stays about the same. A peek into Gale's immaculate lab notebook is one of the highlights here since it reveals the earliest plans for the Superlab and his own methods of meth synthesis; another high point is Walt recruiting the laundry workers to help clean the lab, of course.

Unfortunately, this season also saw the end of the Superlab. After Gus Fring's explosive death (see chapter 8), Walt and Jesse flood the lab with methylamine and flammable solvents while rigging an electrical arc to a timer. This gives the meth-making MacGyvers enough time to flee the laundry facility—and pull the fire alarm—before a raging chemical fire destroys the Superlab for good. It was a beautiful set and the site of quite a bit of fictionalized real-world science. It will be missed.

Cartel Lab

Technically, the Mexican drug cartel's lab is seen while the Superlab is still in operation, but it deserves its own separate subsection since it plays a pivotal role in the evolution of Jesse Pinkman as a meth-making chemist. In Season 4, Episode 10: "Salud," Jesse is escorted to the cartel's chemistry lab by both Gus and Mike under the guise of offering the cartel the man behind the Blue Sky meth. (In actuality, it's part of Gus's elaborate plan for vengeance.) Without his mentor Heisenberg, and on the wrong side of the language barrier, Jesse temporarily finds himself a little out of his element. But all that time spent in the lab under Walt's tutelage, whether it was in the RV or the Superlab, has apparently paid off since Jesse is now confident enough to take the cartel's own lead chemist to task.

Jesse has to perform a cook for the cartel in order to prove his worth, but since the Blue Sky method is still based on the reductive amination process, he'll need PAA as a precursor in order to make his own P2P. The problem is that the cartel doesn't have any on hand ... because they make it themselves. This makes sense considering that the chemical is also under governmental controls and that it's relatively easy to synthesize, especially for a lab set up to make meth in the first place. The phenylacetic acid can be made by hydrolyzing benzyl cyanide, which involves adding water and HCl and kicking out an ammonium chloride leaving group. I don't know how many sophomore chemistry students do this on the regular, but as the cartel's chemist says, they probably *could* do it.

Benzyl cyanide is used as a solvent and as a pharmaceutical precursor. Its usefulness as a precursor for illicit drugs has led to this chemical landing on the DEA's watch list as well, but it can *also* be synthesized from its own precursors. While it's possible to synthesize most of the chemicals needed in the meth-making process from "scratch," the time, cost, and difficulty in doing this is weighed against the associated risks of simply purchasing the necessary precursor chemicals.

Unfortunately for Jesse, he doesn't make his own; he gets his phenylacetic acid "from the barrel with the bee on it." The cartel's chemist oversteps his bounds, however. Jesse's the man that Gus and the cartel brought in to make Blue Sky, so it's his way or the highway. So while one of the cartel chemists is likely assigned to synthesizing Jesse's much-needed PAA, Heisenberg Jr. chastises the rest of the team for the sorry state of the lab, which is in dire need of cleaning and decontamination. Once that's complete, Jesse gets on with the cook. It's the only one we ever get to see in this location. (For a refresher on how his solo cook turned out, revisit chapter 17.)

Vamonos Pest

With both the RV lab and the Superlab destroyed, and their link to the cartel severed, Walt and Jesse are going to need a new place to cook. Enter: Vamonos Pest, introduced in Season 5, Episode 3: "Hazard Pay." Combining the mobility of the RV with the state-of-the-art equipment of the Superlab is this brilliant meth-making setup cooked up by the *Breaking Bad* writers in Season 5. It was Gilligan's idea to move the cook into tented houses—though he later found out that it's pretty rare to tent houses in New Mexico for pest fumigation—and a lot of time went into figuring out how to make that work. Production designer Mark Freeborn put a herculean effort into designing and building the meth-making equipment, making items portable by fitting them in actual road cases that were as big as they possibly could be while still being able to roll through a standard household doorway.

Fun Fact: The color scheme of the Vamonos Pest tents matches the *Breaking Bad* title color scheme.[17]

Another fan-favorite aspect of the show that was introduced in "Hazard Pay" was the new design of Walt and Jesse's "cook suits." The previous suits had proven to be too hot for the actors and so noisy that they interfered with recording dialogue. Season 4's very recognizable, canary yellow Tyvek suits were transformed into Season 5's quieter "track suit"-style attire, though the new suits were still "ungodly hot," as Aaron Paul put it.[18] The show's costume department managed to find a waterproof material that was used in garments; they actually dismantled those garments and used the material to make the new suits for Season 5.[19]

But before the new cook suits could be used—and before settling on Vamonos Pest for their new cover—Saul had to sell Walt, Jesse, and Mike on a new cook site. This "tour de meth" around town took the quartet to various options, including: a box factory, a tortilla-making facility, and a laser tag venue.

The problems with all of these sites? The salt and steam used in the box factory would wreak havoc with the environmental controls of the cook; the food-processing facility would not only invite unannounced government inspections but would also result in tortillas that "smell like cat piss" due to the odors of the chemicals used in cooking (methylamine in particular has a strong fishy smell); and the laser tag front, a running gag in the show, was dismissed by Walt, Jesse, and Mike simply due to personal taste.[20]

Inside *Breaking Bad*: All of the places Saul took Walt, Jesse, and Mike to visit were actually practical locations. The tortilla company was a working facility, requiring the production team to buy out the place for the day due to regulatory and cleanliness issues. While on site they had to wear hairnets all day, but in the end, they had all the tortillas they could eat.

(Aaron Paul also made the mistake of grabbing a tortilla that was fresh out of the oven while filming, burning his mouth in the process, which you can see on camera in the episode.) The box-making factory was inspired by Gilligan's past experience working in a similar facility ... for a grand total of two days.[21]

With Vamonos Pest emerging as the clear winner, the meth-making team set about figuring out the details to make their plan work. Saul takes care of the business and legal sides, and Mike vets employees and beefs up security, while Walt and Jesse design their portable lab. As Walt explains it, a tented house setup for fumigation comes with strange smells and toxic chemicals that act as a "Keep Out" sign for just about everybody, and they stay that way long enough to accommodate a cook without any questions.

The first problem Walt and Jesse tackled is the portability of their lab equipment. Viewers get to see the process unfold through diagrams listing steps in their setup like an "ultra high purity recirc pump, high-pressure extractors, agitator, internal plate column, and low-pressure filtration." Jesse shows some surprising engineering aptitude by suggesting that they make some of their equipment piecemeal, like the finishing tank's agitation motor, which pops on and off the top of the vessel. All of these custom builds offer up a perfect opportunity for savvy junkyard owner Old Joe and his gang to put their TIG welders ("Tungsten Inert Gas" used for welding aluminum) and machining skills to work.

Another upside of using Vamonos Pest as a cover is that no one will look twice when they see containers of toxic chemicals coming off the back of one of their trucks or stored in their warehouse. Just to be safe, Walt and Jesse cover up their barrels of meth chemicals with pesticide labels. Chemicals used by Vamonos Pest include:

Cyfluthrin—A pesticide that's highly toxic to fish, invertebrates, and insects, but less so to humans

Glyphosate—Better known as the active herbicide ingredient in Monsanto's Roundup

Bromadiolone—A rodenticide (or rat poison) known as "super-warfarin" due to its potent anticoagulant properties; it's classified as "extremely hazardous" in the United States.

Chloropicrin—A broad-spectrum antimicrobial/fungicide/herbicide/insecticide/nematicide, regulated as a restricted-use pesticide and used as an indicator "warning" repellent before fumigating with sulfuryl fluoride

Sulfuryl fluoride—An easily condensed, odorless gas used to control termites, though it's also neurotoxic; it's also known by the trade name Vikane

Carbaryl—A popular insecticide used in home gardens.

These guys are not screwing around when it comes to pests.

Fun Fact: When informing the homeowner about the equipment and pesticides they'll be using during the fumigation, the pest control expert drops the term "Blattodea." That's the order of insects containing cockroaches and termites. It was just too fun a word not to include here somewhere. (And there's a hilarious shot of a cockroach scurrying across a countertop during the cook clean-up, by the way.)

In "Hazard Pay," the episode in which this cook site literally unfolds, Jesse can be seen affixing a label of "cyfluthrin" over their barrel of methylamine; a large container bearing the same label is later poured into their reaction vessel, so I hope they have some way of separating the pesticide from their precursor or they're going to be out of business very quickly. Having donned their cook suits, Walt and Jesse enter the smaller tented enclosure inside the already-tented house to kick off yet another beautifully shot, edited, and scored meth-making montage. Viewers get to see how all of the various pieces of equipment unpack from their compact roll-away road cases to fit together quite nicely in the custom mobile lab. It's just brilliant.

It seems like the pair have made some small sacrifices, however. Rather than measuring the pH of their samples using TLC (thin-layer chromatography) like they did in the Superlab, Walt has resorted to the common but less specific practice of using pH indicator strips. The gorgeous, camera-friendly shot of them pouring shredded aluminum foil into a vat is present here as well, more as a delight for the eyes than anything else but also to remind us that they're still using the aluminum amalgam in the reduction step. The use of this particular reduction reagent in the real world results in the appearance of cloudy gray foam, which is captured nicely in this montage.[22] Then, it's all over but the waiting, so what better way to pass the time than by watching *The Three Stooges* right in the comfort of someone else's living room?

As is the case in *Breaking Bad*, everything goes along just fine for the next few episodes, but by Season 5, Episode 7: "Say My Name," things start to go south once more. Mike is done with Walt's scheming and his increasingly violent alter ego, Heisenberg. Jesse, finally listening to Mike's earnest advice, is on his way out, too, despite the manipulative Walt saying that Jesse's every bit as good a cook as he is. But with a new production and distribution deal made with Arizona-based competitor Declan, Heisenberg is going to need an assistant if he's to stay on schedule. Enter Todd, the intrepid go-getter who's willing to do whatever it takes to get ahead.

With Todd as his assistant, Walt is starting from scratch. At least Jesse had taken high school chemistry, even if he failed at it. Walt recognizes this fact, saying that he doesn't need Todd to be Antoine Lavoisier, a reference to the "father of modern chemistry." The eighteenth-century French chemist who heavily influenced the future of both chemistry and biology was known for implementing a quantitative approach to science in place of the commonly used qualitative one. (Lavoisier also discovered oxygen's role in combustion, naming and recognizing both oxygen and hydrogen as the gases involved in the process rather than subscribing to the prevailing so-called phlogiston theory of his time that proposed a "fire-like" element. He also constructed the metric system and refined chemical nomenclature, to name but a few of his accomplishments.) Todd, no Lavoisier, is just trying to keep up with Walt.

During the cook, specifically while dumping the aluminum foil shreds into a reaction vessel, Walt mentions to Todd that "aluminum helps speed the delivery of the hydrogen chloride." If you've been following along, you'll remember that aluminum, when added to a mercury solution, forms an amalgam that reduces the imine to the final amine, methamphetamine. However, the only way the aluminum helps here is if using it gets Walt and Todd through the reaction more quickly so that *they* can "deliver" the hydrogen chloride during the crystallization step. Thanks for keeping it simple, Walt.

At the end of the cook, Walt's final bit of instruction to Todd is that "the CO_2 freezes the liquid, crystallizing it," which gives them their crystal methamphetamine. Curiously, because Heisenberg's meth is so pure, it will actually crystallize more slowly than "dirty meth" that has more impurities and contaminants. Crystals need a nucleation site, or a "seed," in order to have an anchor point from which to grow, similar to how snow forms as accumulations of ice crystals in the atmosphere; those impurities and contaminants provide those seeds. Walt may be trying to speed up crystallization of his methamphetamine not only by dropping the temperature but also by using CO_2 to keep water vapor out of the equation. The "ice" he's looking for isn't water ice, after all.

A few more meth-making montages focus on Todd trying to get the process right, but like the labs before it, Vamonos Pest isn't long for the world of *Breaking Bad*. Things take a dark and dirty turn from here on out, so hold tight as we take a look at the even gnarlier side of the illicit methamphetamine business.

Declan's Desert Lab

By the second half of Season 5, Walt is out of the drug business, partly because he's made more money than he knows what to do with while eliminating just about everyone who could implicate him along the way, and partly because his cancer is back and he has only a few months left to live. But just because Heisenberg has retired doesn't mean his suppliers, buyers, and competition will do the same. One of the labs still in operation—briefly—is Declan's desert facility seen in Season 5, Episode 10: "Buried." Do yourself a favor and revisit the scene where Lydia Rodarte-Quayle descends into this dark and dingy place. The production design and construction crews did a bang-up job on this build, which is supposed to be a full bus buried out in the desert, outfitted as a clandestine lab.

While the exterior location was simply a hole dug about three feet down into the ground to provide the illusion of someone lowering themselves down into the buried facility, the practical set of the lab itself was a Frankenstein's monster of a soundstage put together by the show's electrical department and grips. There was a full working framework built around the bus to make it easier and more efficient for the camera crew to work. The bus itself was also cut into sections that were put on rollers, making it easier to move separate parts of it for filming. It's another fantastic design element that was seen only too briefly on the show.[23]

For the purpose of the show's narrative, however, we never even get to see a meth-making session here. Declan's cook's substandard product—measuring only 68 percent purity—is untenable. (Even Todd's cooks came in at 74 percent before he nearly burned the lab down.) After Declan and his team are eliminated on Lydia's orders, the new meth-backed power players in town are Jack Welker—Todd's uncle—and his White Supremacist Gang. They scavenge the subterranean meth lab for equipment and supplies in order to get everything they need for their new cook: Todd Alquist.

Jack Welker's White Supremacist Compound

This spacious, well-appointed Quonset hut is where it all comes to a head as the season and the series prepare to end. In Season 5, Episode 13: "To'hajiilee," viewers get their first eyeful of Todd's cook at the other desert compound, the one presided over by a bunch of no-good Nazis. This cook just isn't pretty. The meth produced may be closer to what real-world meth actually looks like—a beige, somewhat cloudy color—but it sure does make you miss the sight of Blue Sky. Todd's yield is pretty good here, around fifty pounds of product or so, even if the appearance and the purity—76 percent, which is better than his previous cooks, at least—are subpar. What's interesting in this session is the manner in which Todd measures that percentage.

Sharp-eyed viewers will spot the handheld device Todd uses here. He places a sample of the liquid meth onto a small window on one end of the device while looking through an eyepiece on the other end. Satisfied that he saw whatever it was he needed to see, Todd checks what I presume to be a list of values, and then proudly states that his meth is 76 percent pure. (Still not blue though ...) So what's going on here, exactly?

Well, it looks like Todd's using a refractometer. This is a device used in the lab or even the field to measure an index of refraction, basically a measure of how light moves through a given medium. It can be used to measure plasma protein in blood samples in veterinary applications, specific gravity of urine in drug-testing scenarios, and even in home brewing and beekeeping to monitor sugar and water content, respectively. In this case, it's a measure of how light moves through the meth sample. A denser sample that's more packed full of dissolved solids will slow and scatter the passage of light through it, relative to a purer sample that's less dense and optically clear. If Todd is comparing his reading to a table of known refractive indices for methamphetamines of varying purity, this is actually a decent quick-and-dirty method of measuring that. It does nothing to tell Todd which impurities are present, chemically, but it's an interesting addition nonetheless.

Todd, however, isn't quite cutting it as far as Lydia and her Czech Republic buyers are concerned. But the good news for them, and the bad news for Jesse, is that Walt's original apprentice is still around. In Season 5, Episode 14: "Ozymandias," Jack's gang has a special pit at the compound where they keep Jesse prisoner, only letting him out to cook for them while keeping him tethered like an animal. It's a tough thing to revisit here, and even tougher to watch, but hang in there because we're almost done!

The lab, which used to be a place of refuge and even enjoyment for Jesse, has now become a living Hell. He even resorts to living in his memories of long ago, before drugs changed his life, a time when he was able to find simple pleasures in the noble art of woodworking. It's a hard cut from that warm, sepia-toned flashback to the cold, sterile, and metallic environment of the prison lab. We get to see Jesse prepare to dump the shredded aluminum foil one last time and it's just the saddest thing in the world, or at least arguably one of the saddest things that happens on *Breaking Bad*.

Since the show itself started in the lab way back in the pilot, it's only fitting that it ends there in the series finale, "Felina." Having taken out the Nazis and at long last set Jesse free, Walter White, the man infamously known as Heisenberg, all alone now, estranged from his family, bids a fond farewell to the equipment he'd spent so much time with. He succumbs to his gunshot wound and dies in the only place he ever truly felt alive: a clandestine methamphetamine lab in the lawless desert of the American Southwest.

Side RxN #14: Qualitative Analysis of Foods at Madrigal

You didn't really think I'd end this section on such a down note, did you? Not when there are varieties of dipping sauces to talk about! In Season 5, Episode 2: "Madrigal," Peter Schuler, head of the fast-food division of Madrigal Electromotive GmbH, which owns Gus Fring's Los Pollos Hermanos chain, is sampling new flavors of dipping sauce in a state-of-the-art laboratory with a bowl full of chicken nuggets at his disposal. As delightful as this sounds as a paying gig, Herr Schuler is not enjoying himself. Let's take a look at his flavorful options, shall we?

Honey Mustard—A sweeter variety with a Brix number increased by 14 percent by using high-fructose corn syrup and 2.2 percent less honey.

Franch—Half French dressing, half ranch. A new conception.

Cajun Kick-Ass—Since the original caused gastric distress, this is a reformulation.

Smoky Mesquite BBQ—Now with 3 percent more smoke flavor.

Ketchup—The old standby.

As silly as this all sounds, there's some science at work here, too. The Brix number, or degrees Brix (°Bx), is a measure of the sugar content of a solution. One degree Brix equals 1g of sucrose in 100g of solution, so it basically represents the percentage of sugar present. If the original Honey Mustard dipping sauce measured, say, 30°Bx, then the sweetened "American Midwest" version would be 34.2°Bx representing a 14 percent increase.

The value was named for nineteenth-century German mathematician and engineer Adolf Ferdinand Wenceslaus Brix, one of a few men who busied themselves dabbling in measuring specific gravity of sucrose solutions in the 1800s. (Simpler times.) Others include Karl Balling, who established the Balling scale (°Balling) to measure the concentration of dissolved solids in brewery wort as a potential indicator of its alcoholic strength, and Fritz Plato, whose degree Plato (°P) is also widely used in the brewing industry for the same purpose. I mention all of this in part because the determination of a solution's sugar content can also be done by measuring its refractive index, much like Todd did while measuring his meth sample. The other reason I mention it is because this scene is just plain silly.

Fun Fact: You can actually be gainfully employed as a sensory analyst. More than just a "taste tester," sensory experts are expected to have a palate that can detect a wide variety of chemical compounds (aromas and flavors) as well as the knowledge, ability, and experience to be able to communicate those findings clearly. Also, analysts are expected to sample not just the final product, but all of the ingredients that go into the process as well. A brewer worth her hops will not only sample the bottled brew, but also the barley, malt, hops, water, and wort along the way.

Not-So-Fun Fact: The distressed Schuler excuses himself from the tasting session and bypasses the law enforcement team that has just arrived to take him into custody for questioning. Preferring oblivion to incarceration, Herr Schuler uses an automated external defibrillator (AED) to commit suicide. The bad news is that this section has become a downer once again; the good news is that this probably wouldn't have worked in the real world.

Since AEDs are "dummy-proofed" to allow operation by untrained individuals out in the world, they'll only work under certain conditions, namely if the chest pads are properly placed and if the patient's heart is found to be in fibrillation, meaning, in an irregular rhythm. The amount of energy provided by an AED is on the order of 120–200 joules up to three times, which is aimed at correcting arrhythmia in the heart. Don't confuse these with manual defibrillators, however, which should only be left to medical professionals and those who play them on TV.

XIX Finale/Felina

From Dr. Donna J. Nelson:

Breaking Bad has used many iconic images as symbols for the show, and these have changed as the show progressed. The first was the set of periodic table symbols for bromine and barium, joined diagonally at the corners, and printed on almost everything. I was given refrigerator magnets with the symbols.

During my set visits in Seasons 1 and 4, I received commemorative coins that were given to the crew (see figures 19.1 and 19.2). These coins bore symbols commonly used for the show. The obverse of the Season 1 coin displayed one of the first symbols used—Walter White's "tighty whitey" underwear, along with the words "BREAKING BAD" and "IT'S CHEMISTRY, BITCH." The reverse had the symbols for bromine and barium joined at the corners, with the words "SEASON ONE CREW * ALBUQUERQUE * 2007."

The Season 4 coin's obverse had the classic brown felt men's pork pie hat from the 1940s, which Walter White wore when he appeared as Heisenberg, whose persona became associated with the hat. It also displayed the words "BREAKING BAD" and "HAT ON, GLOVES OFF." The reverse bore the symbols for bromine

Figure 19.1
Breaking Bad commemorative coins. Image courtesy of Dr. Donna J. Nelson.

Figure 19.2
Breaking Bad commemorative coins. Image courtesy of Dr. Donna J. Nelson.

and barium joined at the corners, with the words "SEASON FOUR CREW * ALBUQUERQUE BURBANK * 2011."

Blue crystals of methamphetamine became one of the most recognizable and popular images. At the 2012 Comic Con, the *Breaking Bad* actors and crew threw small packets to excited attendees. These packets contained blue crystals, which were actually sugar crystals, flavored to taste like cotton candy.

During one of my set visits, Vince asked me "What do you think about making the meth blue?" I said, "I wouldn't do it. Methamphetamine is white." He said, "We have pure meth; is there a reason why pure meth might be blue?" I said, "Pure meth will be white." He asked, "What if it is really, really pure meth?" I replied, "Really, really pure meth will be really, really white." However, as we all know, he ignored my advice and made the meth blue. As I recall, this is the only time he didn't take my advice. However, this show is not a documentary; it is fiction. The blue meth named "Blue Sky" was a plot device and a symbol. Walt needed a trademark for his product, and this was its color.

Other symbols are Walt's facial expression, his yellow hazmat suit, and some of his remarks, such as "Remember my name." This last one is also a plot device, because it encourages us to remember the name of the show, *Breaking Bad*, and many of us will remember it.

Cheer up, beautiful people. This is where you get to make it right.—Walter White, Season 5, Episode 16: "Felina"

Over the course of five seasons of *Breaking Bad*, audiences experienced Walter White's transformation from a mild-mannered milquetoast, who was struggling to make ends meet as a high school science teacher, into Heisenberg, a power-hungry drug kingpin whose genius was only outmatched by his ruthlessness. This change from zero to anti-hero to outright villain was complemented every step of the way by Walt's scientific knowledge and understanding, which allowed him not only to build his meth empire from the lab up, but also to remove any obstacles—organic or inorganic—that stood in his way. And in the series finale, "Felina," there is one last loose end to tie up.

There's not much science on display in this episode (though I do like the "transformation" of the word "Felina" into "Finale," and vice versa) since it centers on Walt putting his final plan into action before shuffling off this mortal coil, but there's one particularly impressive bit of production design I wanted to highlight.

In the Season 5 premiere, "Live Free or Die," Walt reveals to the audience that he has an M60 machine gun casually stowed away in the trunk of his car alongside boxes of ammunition. The specter of it hangs over the rest of the season until the final few moments of the finale. In spectacular fashion, it's revealed that Walt has mounted the M60 to some sort of automatic turret contraption bolted into the trunk of his car and activated by his remote key fob. The business end of the machine gun takes out the white supremacists in Jack's hideout and also mortally wounds Walt himself.

While this is more MacGyver than mad science, the production design behind the weapon's setup is rather clever. Originally, Gilligan and the writers referred to this as "Satan's windshield wiper" in the episode's script.[1] As visually evocative as that description is, it's also partially literal: Gilligan had conceived of attaching the M60 to a windshield wiper motor to give the gun its deadly oscillating, sweeping ability. His production team stepped in and beefed up the rig by using a garage door motor instead, giving the contraption enough oomph to oscillate the twenty-three-pound gun. According to Gilligan, while speaking on the episode's related behind-the-scenes podcast, the motor was a 12V garage door opener powered by a standard 12V car battery.[2] Normally, however, the garage door motor is plugged into the 120V household power, but a 12V battery backup can be used to operate the motor during power outages; in other words, this checks out.

As for whether or not this setup would actually work, and whether or not it would have deadly consequences for the Nazis on the other side of the wall, luckily, the *MythBusters* team put this one to the test.[3] While it's not surprising that their test looked almost exactly like the scene from the finale, what *is* surprising is that the *MythBusters'* stand-in for a prone Walter White actually managed to survive the assault.

But in the reality of *Breaking Bad,* Walt's story came to an end shortly after his final act of mad science. His transformation was complete, and after two years, his reaction had run its course. I find it endlessly fascinating that all the traits that made Heisenberg such a dangerous and deadly adversary were part of Walter White's makeup all along. We don't know every detail about the fifty years that preceded his cancer diagnosis and his ultimate decision to break bad, but we can be confident in saying that Walt wasn't a reformed criminal sliding back into his evil ways, nor was he some sort of overnight supervillain. He was simply a man, a man gifted with extraordinary scientific knowledge and insight, but a man just the same. Heisenberg was the darkness inside of Walter White that simmered beneath the surface for fifty years before finally boiling over as an unquenchable, runaway reaction, catalyzed by his cancer diagnosis. As viewers, we can choose (and argue over) the moment that Walter White truly becomes a villain, but for Heisenberg himself, stopping his ascension to drug kingpin or descent to villainy was never an option.

That element of *Breaking Bad,* the spectrum of morality that exists to allow individual viewers to decide for themselves just when and where Walt has gone too far, is one of the many things that make this drama one of the best in television history. Millions of viewers, numerous awards, and heaps of critical acclaim offer plenty of evidence to that fact. But *Breaking Bad* also remains a rich and rewarding form of personal entertainment. You can analyze the show from any number of perspectives: historical or legal accuracy, or both, the phenomenal musical work of show composer Dave Porter, the incredible achievements of the show's production team and art department, the vital contributions of the cast and crew, a critical approach to the show itself and how it's changed the television landscape, or even from the visual perspective of color theory. (I encourage you to read all the available material written on these topics, and if it doesn't exist, I further encourage you to do your own analysis!)

Then, of course, you can take a look at the slice of the *Breaking Bad* pie that has to do with real-world, practical science and how accurate the show's depiction really is. If you've made it this far, I'd say you can consider yourself an expert in this regard.

Breaking Bad will remain one of the greatest shows ever to grace the small screen. The fact that it's one of the very few fictional dramas to embrace real scientific concepts and practices is just the icing on the cake. It's been a privilege and a pleasure to revisit them with you. I hope that in doing so you've found an even deeper appreciation for real-world science (and scientists) and its integral presence on the show, and that you've enjoyed reading along with *The Science of "Breaking Bad."*

Glossary

Acid (n.) A substance with particular chemical properties including turning litmus red, neutralizing alkalis, and dissolving some metals; typically, a corrosive or sour-tasting liquid of this kind. Chemistry—A molecule or other species that can donate a proton or accept an electron pair in reactions.

Adenocarcinoma (n.) Medicine—A malignant tumor formed from glandular structures in epithelial tissue.

Addiction (n.) The fact or condition of being addicted to a particular substance or activity. Addiction is a primary, chronic disease of brain reward, motivation, memory, and related circuitry. Addiction is characterized by the inability to consistently abstain, impairment in behavioral control, craving, diminished recognition of significant problems with one's behaviors and interpersonal relationships, and a dysfunctional emotional response.

Allotrope (n.) Chemistry—Each of two or more different physical forms in which an element can exist. Graphite, charcoal, and diamond are all allotropes of carbon.

Anode (n.) The positively charged electrode by which the electrons leave an electrical device. The opposite of cathode.

Atom (n.) The smallest particle of a chemical element that can exist.

Atomic number (n.) Chemistry—The number of protons in the nucleus of an atom, which is characteristic of a chemical element and determines its place in the periodic table.

Autoclave (n.) A strong heated container used for chemical reactions and other processes using high pressures and temperatures, such as steam sterilization.

Base (n.) Chemistry—A substance capable of reacting with an acid to form a salt and water, or (more broadly) of accepting or neutralizing hydrogen ions.

Battery (n.) A container consisting of one or more cells, in which chemical energy is converted into electricity and used as a source of power.

Cathode (n.) The negatively charged electrode by which electrons enter an electrical device. The opposite of anode.

Chemistry (n.) The branch of science concerned with the substances of which matter is composed, the investigation of their properties and reactions, and the use of such reactions to form new substances.

Chiral (adj.) Chemistry—Asymmetric in such a way that the structure and its mirror image are not superimposable.

Conjugate (adj.) Chemistry—(of an acid or base) related to the corresponding base or acid by loss or gain of a proton.

Dilution (n.) The action of making a liquid more dilute.

Dissociation constant (n.) Chemistry—A quantity expressing the extent to which a particular substance in solution is dissociated into ions, equal to the product of the concentrations of the respective ions divided by the concentration of the undissociated molecule.

Drug (n.) A medicine or other substance that has a physiological effect when ingested or otherwise introduced into the body. A substance taken for its narcotic or stimulant effects, often illegally.

Electron (n.) Physics—A stable subatomic particle with a charge of negative electricity, found in all atoms and acting as the primary carrier of electricity in solids.

Electronegativity (n.) Chemistry—The degree to which an element tends to gain electrons and form negative ions in chemical reactions.

Element (n.) Each of more than one hundred substances that cannot be chemically interconverted or broken down into simpler substances and are primary constituents of matter. Each element is distinguished by its atomic number, meaning the number of protons in the nuclei of its atoms.

Enantiomer (n.) Chemistry—Each of a pair of molecules that are mirror images of each other.

Endothermic (adj.) Chemistry—(of a reaction or process) accompanied by or requiring the absorption of heat; (of a compound) requiring a net input of heat for its formation from its constituent elements. The opposite of exothermic.

Enthalpy (n.) Physics—A thermodynamic quantity equivalent to the total heat content of a system. It is equal to the internal energy of the system plus the product of pressure and volume; the change in enthalpy associated with a particular chemical process.

Equilibrium (n.) Chemistry—A state in which a process and its reverse are occurring at equal rates so that no overall change is taking place.

Exothermic reaction (n.) Chemistry—(of a reaction or process) accompanied by the release of heat; (of a compound) formed from its constituent elements with a net release of heat. The opposite of endothermic.

Frequency (n.) The rate per second of a vibration constituting a wave, either in a material (as in sound waves), or in an electromagnetic field (as in radio waves and light).

Fugue (n.) Psychiatry—A loss of awareness of one's identity, often coupled with flight from one's usual environment, associated with certain forms of hysteria and epilepsy.

Galvanic/voltaic (adj.) Relating to electricity produced by chemical action in a battery.

Heisenberg, Werner Karl (1901–1976), German mathematical physicist and philosopher. He developed a system of quantum mechanics based on matrix algebra in which he stated his famous uncertainty principle (1927). For this and his discovery of the allotropic forms of hydrogen he was awarded the 1932 Nobel Prize for Physics.

Isomer (n.) Chemistry—Each of two or more compounds with the same formula but a different arrangement of atoms in the molecule and different properties. Physics—Each of two or more atomic nuclei that have the same atomic number and the same mass number but different energy states.

Isotope (n.) Chemistry—Each of two or more forms of the same element that contain equal numbers of protons but different numbers of neutrons in their nuclei, and hence differ in relative atomic mass but not in chemical properties; in particular, a radioactive form of an element.

Methamphetamine (n.) A synthetic drug with more rapid and lasting effects than amphetamine, used illegally as a stimulant.

Molar (n.) Relating to mass; acting on or by means of large masses or units. Chemistry—Relating to one mole of a substance; (of a solution) containing one mole of solute per liter of solution.

Neurotransmitter (n.) Physiology—A chemical substance that is released at the end of a nerve fiber by the arrival of a nerve impulse and, by diffusing across the synapse or junction, effects the transfer of the impulse to another nerve fiber, a muscle fiber, or some other structure.

Nucleus (n.) The positively charged central core of an atom, consisting of protons and neutrons and containing nearly all its mass.

Oxidation number/state (n.) Chemistry—A number assigned to an element in chemical combination that represents the number of electrons lost (or gained, if the number is negative), by an atom of that element in the compound.

Panic attack (n.) A sudden overwhelming feeling of acute and disabling anxiety.

Periodic table of elements (n.) Chemistry—A table of the chemical elements arranged in order of atomic number, usually in rows, so that elements with similar atomic structure (and hence similar chemical properties) appear in vertical columns.

Photon (n.) Physics—A particle representing a quantum of light or other electromagnetic radiation. A photon carries energy proportional to the radiation frequency but has zero rest mass.

Planck's constant (n.) Physics—A fundamental constant, equal to the energy of a quantum of electromagnetic radiation divided by its frequency, with a value of 6.626×10^{-34} joule-seconds.

Post-traumatic stress disorder (PTSD) (n.) Medicine—A condition of persistent mental and emotional stress occurring as a result of injury or severe psychological shock, typically involving disturbance of sleep and constant vivid recall of the experience, with dulled responses to others and to the outside world.

Precursor (n.) A substance from which another is formed, especially by metabolic reaction.

Proton (n.) Physics—A stable subatomic particle occurring in all atomic nuclei, with a positive electric charge equal in magnitude to that of an electron.

Redox (n.) Chemistry—Oxidation and reduction considered together as complementary processes.

Reduction (n.) Chemistry—The process or result of reducing or being reduced; the gain of electrons and decreasing of the oxidation state.

Oxidation (n.) Chemistry—The process or result of oxidizing or being oxidized; the loss of electrons and increasing of the oxidation state.

Salt bridge (n.) Chemistry—A tube containing an electrolyte (typically in the form of a gel), providing electrical contact between two solutions.

Standard atomic weight/relative atomic mass (n.) Chemistry—The ratio of the average mass of one atom of an element to one twelfth of the mass of an atom of carbon-12.

Tannin (n.) A yellowish or brownish bitter-tasting organic substance present in some galls, barks, and other plant tissues, consisting of derivatives of gallic acid.

Vapor Pressure (n.) Chemistry—The pressure of a vapor in contact with its liquid or solid form.

Wavelength (n.) Physics—The distance between successive crests of a wave, especially points in a sound wave or electromagnetic wave.

Definitions via https://en.oxforddictionaries.com, accessed November 9, 2018.

Notes

Chapter I

1. Kelley Dixon and Vince Gilligan, "Episode 1," *Breaking Bad Insider Podcast,* AMC, March 26, 2009, https://www.stitcher.com/podcast/breaking-bad-insider-podcast/e/38708099.

2. Ibid.

3. Jyllian Kemsley, "'Breaking Bad': Novel TV Show Features Chemist Making Crystal Meth," *Chemical & Engineering News* 86, no. 9 (March 2008): 32–33, https://cen.acs.org/articles/86/i9/Breaking-Bad.html.

Chapter II

1. Katrina Krämer, "Beyond Element 118: The Next Row of the Periodic Table," *Chemistry World,* January 29, 2016, https://www.chemistryworld.com/news/beyond-element-118-the-next-row-of-the-periodic-table/9400.article.

2. "Unified Atomic Mass Unit" defined, *IUPAC Gold Book,* last update February 24, 2014, http://goldbook.iupac.org/html/U/U06554.html.

Chapter III

1. "The Science of *Breaking Bad*: Pilot," *Weak Interactions: Screen Science Explained,* accessed November 13, 2018, https://weakinteractions.wordpress.com/2009/06/15/the-science-of-breaking-bad-pilot/.

2. Anne Marie Helmenstine, "Colored Fire Spray Bottles," ThoughtCo, last updated June 8, 2018, https://www.thoughtco.com/colored-fire-spray-bottles-607497; "Pyrotechnic compounds," under "Fireworks," last updated November 3, 2018, accessed November 16, 2018, https://en.wikipedia.org/wiki/Fireworks#Pyrotechnic_compounds.

3. David Harvey, "10.7: Atomic Emission Spectroscopy," last updated May 3, 2016, https://chem.libretexts.org/Textbook_Maps/Analytical_Chemistry_Textbook_Maps/Map%3A_Analytical_Chemistry_2.0_(Harvey)/10_Spectroscopic_Methods/10.7%3A_Atomic_Emission_Spectroscopy.

Chapter IV

1. Matheson Tri-Gas, *Material Safety Data Sheet: Phosphine,* April 18, 2006, https://www.mathesongas.com/pdfs/msds/MATH0083.pdf

2. Royal Society of Chemistry, "Periodic Table: Phosphorus," accessed October 28, 2018, http://www.rsc.org/periodic-table/element/15/phosphorus.

3. Jonathan Hare, "On-Screen Chemistry," *InfoChem,* accessed October 28, 2018, http://www.rsc.org/images/breaking-bad-phosphine-gas_tcm18-233821.pdf.

4. "Red Phosphorus Flame Retardants," SpecialChem, accessed October 28, 2018, http://polymer-additives.specialchem.com/selection-guide/flame-retardants-center/red-phosphorus.

5. Juliet Lapidos, "What's So Great about White Phosphorus?," *Slate,* March 27, 2009, http://www.slate.com/articles/news_and_politics/explainer/2009/03/whats_so_great_about_white_phosphorus.html.

6. Cameo Chemicals, "Chemical Datasheet: Calcium Phosphide," accessed October 28, 2018, https://cameochemicals.noaa.gov/chemical/314.

7. Hare, "On-Screen Chemistry."

8. National Research Council, *Toxicity of Military Smokes and Obscurants: Volume 1* (Washington, DC: National Academies Press, 1997), https://www.nap.edu/read/5582/chapter/6#99.

9. Jeremy Pearce, "H. Tracy Hall, a Maker of Diamonds, Dies at 88," *New York Times,* August 2, 2008, http://www.nytimes.com/2008/08/02/us/02hall.html.

10. Thomas H. Maugh II, "General Electric Chemist Invented Process for Making Diamonds in Lab," *Los Angeles Times,* July 31, 2008, http://articles.latimes.com/2008/jul/31/local/me-hall31.

Chapter V

1. Kelley Dixon and Vince Gilligan, "Episode 307," *Breaking Bad Insider Podcast,* AMC, May 4, 2010, https://www.stitcher.com/podcast/breaking-bad-insider-podcast/e/38708187.

2. Kelley Dixon and Vince Gilligan, "Episode 9," *Breaking Bad Insider Podcast*, AMC, May 4, 2009, https://www.stitcher.com/podcast/breaking-bad-insider-podcast/e/38708144.

3. Laura June, "Yissum Develops Potato-Powered Batteries for the Developing World," *Engadget*, June 20, 2010, https://www.engadget.com/2010/06/20/yissum-develops-potato-powered-batteries-for-the-developing-worl/.

4. "Standard Electrode Potential," last edited October 23, 2018, accessed October 28, 2018, https://en.wikipedia.org/wiki/Standard_electrode_potential.

5. Ibid.

6. "Peak Amps vs Cranking Amps (and More)," Jumpstarter, accessed October 28, 2018, https://jumpstarter.io/peak-amps-vs-cranking-amps/; Alice Vincent, "*Breaking Bad*: The Science Behind the Fiction," *The Telegraph,* August 3, 2013, http://www.telegraph.co.uk/culture/tvandradio/10218885/Breaking-Bad-The-science-behind-the-fiction.html.

7. "IMERC Fact Sheet: Mercury Use in Batteries," Northeast Waste Management Officials' Association, last updated January 2010, http://www.newmoa.org/prevention/mercury/imerc/factsheets/batteries.cfm.

8. Kelley Dixon and Vince Gilligan, "Episode 313," *Breaking Bad Insider Podcast*, AMC, June 15, 2010, https://www.stitcher.com/podcast/breaking-bad-insider-podcast/e/38708217.

9. Pacific Gas and Electric Company, *PG&E Urges Customers to Keep Metallic Balloons Secure for Valentine's Day Celebrations*, February 6, 2014, https://www.pge.com/about/newsroom/newsreleases/20140206/pge_urges_customers_to_keep_metallic_balloons_secure_for_valentines_day_celebrations.shtml.

10. Dixon and Gilligan, "Episode 313."

Chapter VI

1. "LM Fabricated Series Scrap Magnets," Walker Magnetics, accessed October 18, 2018, http://www.walkermagnet.com/scrap-magnets-liftmaster-lm-fab-series.htm.

2. "Hard Drive Destruction," K&J Magnetics, Inc., accessed October 18, 2018, http://www.kjmagnetics.com/blog.asp?p=hard-drive-destruction.

3. "Common Degausser Misconceptions," Data Security, Inc., accessed October 18, 2018, http://datasecurityinc.com/degaussermyths.html.

4. Kelley Dixon and Vince Gilligan, "Episode 501," *Breaking Bad Insider Podcast*, AMC, July 17, 2012, https://www.stitcher.com/podcast/breaking-bad-insider-podcast/e/38708278.

5. Dave Itzkoff, "Creative Abetting of a TV Drug Lord," *The New York Times*, July 16, 2012, http://www.nytimes.com/2012/07/16/arts/television/breaking-bad-creating-magnetic-attraction.html.

6. Andrew McHutchon, *Electromagnetism Laws and Equations*, 2013, http://mlg.eng.cam.ac.uk/mchutchon/electromagnetismeqns.pdf.

7. AMC, *Making of the Season 5 Premiere: Inside Breaking Bad*, July 15, 2012, https://www.youtube.com/watch?v=01oAqfozRPc.

8. Kelley Dixon and Vince Gilligan, "Episode 12," *Breaking Bad Insider Podcast*, AMC, May 25, 2009, https://www.stitcher.com/podcast/breaking-bad-insider-podcast/e/38708153.

9. Ryan Singel, "Zombie Computers Decried as Imminent National Threat," *Wired*, April 9, 2008, https://www.wired.com/2008/04/zombie-computer/.

10. Kris Hundley and Kendall Taggart, "'Breaking Bad' Fundraiser Funnels Cash to One of America's Worst Charities," Tampa Bay Times, last updated August 22, 2013, https://www.tampabay.com/news/business/breaking-bad-fundraiser-funnels-cash-to-one-of-americas-worst-charities/2137747.

Chapter VII

1. Kelley Dixon and Vince Gilligan, "Episode 308," *Breaking Bad Insider Podcast*, AMC, May 1, 2010, https://www.stitcher.com/podcast/breaking-bad-insider-podcast/e/38708192.

2. Kelley Dixon and Vince Gilligan, "Episode 503," *Breaking Bad Insider Podcast*, AMC, July 31, 2012, https://www.stitcher.com/podcast/breaking-bad-insider-podcast/e/38708283.

3. "Coffee Chemistry: Cause of Bitter Coffee," Coffee Research Institute, accessed October 28, 2018, http://www.coffeeresearch.org/science/bittermain.htm.

4. "How to Brew Coffee", National Coffee Association, accessed October 28, 2018, http://www.ncausa.org/About-Coffee/How-to-Brew-Coffee.

5. Rebecca Smith, "Secret to Perfect Cup of Coffee Lies in the Quality of the Water Researchers Say," *The Telegraph,* June 5, 2014, http://www.telegraph.co.uk/news/health/news/10875537/Secret-to-perfect-cup-of-coffee-lies-in-the-quality-of-the-water-researchers-say.html.

6. Paul B. Schwartz, Samuel Beattie, and Howard H. Casper, "Relationship Between *Fusarium* Infestation of Barley and the Gushing Potential of Malt," *Journal of the Institute of Brewing* 102 (March–April 1996): 93–96, http://onlinelibrary.wiley.com/doi/10.1002/j.2050-0416.1996.tb00899.x/epdf.

7. "Beer Priming Calculator," Brewer's Friend, last updated July 2013, https://www.brewersfriend.com/beer-priming-calculator/.

8. Kelley Dixon and Vince Gilligan, "Episode 5," *Breaking Bad Insider Podcast*, AMC, April 8, 2009, https://www.stitcher.com/podcast/breaking-bad-insider-podcast/e/38708123.

9. "Q&A—Dean Norris (Hank Schrader)," AMC, September 2011, http://www.amc.com/shows/breaking-bad/talk/2011/09/dean-norris-interview-2.

Chapter VIII

1. "628-86-4 (Mercury Fulminate, Wetted with Not Less Than 20% Water, or Mixture of Alcohol and Water, by Mass) Product Description," Chemical Book, accessed October 28, 2018, http://www.chemicalbook.com/ChemicalProductProperty_US_CB6852038.aspx

2. "MythBusters Episode 206: Breaking Bad Special," MythBusters Results, accessed November 13, 2018, https://mythresults.com/breaking-bad-special.

3. Bruce A. Averill and Patricia Eldredge, "8.5 Lewis Structures and Covalent Bonding," in *Principles of General Chemistry* (Creative Commons, December 29, 2012), https://2012books.lardbucket.org/books/principles-of-general-chemistry-v1.0/s12-05-lewis-structures-and-covalent-.html.

4. Edward Howard, "On a New Fulminating Mercury," *Philosophical Transactions* 90 (January 1800), 204–238, http://rstl.royalsocietypublishing.org/content/90/204.full.pdf+html.

5. "MythBusters Episode 206: Breaking Bad Special."

6. Ibid.

7. Howard, "On a New Fulminating Mercury," 206.

8. "MythBusters Episode 206: Breaking Bad Special."

9. Ibid.

10. Kelley Dixon and Vince Gilligan, "Episode 3," *Breaking Bad Insider Podcast*, AMC, March 26, 2009, https://www.stitcher.com/podcast/breaking-bad-insider-podcast/e/38708113; *"Turning the Tables: Inside Breaking Bad—Season 1, Episode 6,"* AMC, accessed October 28, 2018, http://www.amc.com/shows/breaking-bad/video-extras/season-01/episode-06/turning-the-tables-inside-breaking-bad.

11. Kelley Dixon and Vince Gilligan, "Episode 413," *Breaking Bad Insider Podcast*, AMC, October 11, 2011, https://www.stitcher.com/podcast/breaking-bad-insider-podcast/e/38708275.

12. Kelley Dixon and Vince Gilligan, "Episode 412," *Breaking Bad Insider Podcast*, AMC, October 4, 2011, https://www.stitcher.com/podcast/breaking-bad-insider-podcast/e/38708270.

13. "Ammonium Nitrate Security Program (ANSP)," Department of Homeland Security, last updated August 22, 2018, https://www.dhs.gov/ammonium-nitrate-security-program.

14. Dixon and Gilligan, "Episode 413."

Chapter IX

1. Hans Goldschmidt and Claude Vautin, "Aluminum as a Heating and Reducing Agent," *The Journal of the Society of Chemical Industry* 6, no. 17 (June 1898): 543–545, https://web.archive.org/web/20110715133307/http://www.pyrobin.com/files/thermit%28e%29%20journal.pdf/.

2. Signe Brewster, "*Breaking Bad*'s Science Advisor Fact Checks Some of the Show's Greatest Chemistry Moments," Gigaom, August 11, 2013, https://gigaom.com/2013/08/11/breaking-bads-science-advisor-fact-checks-some-of-the-shows-greatest-chemistry-moments/.

3. James Simpson, "A Nazi War Train Hauled the Biggest Gun Ever Made," *Medium*, July 31, 2015, https://medium.com/war-is-boring/a-nazi-war-train-hauled-the-biggest-gun-ever-made-a05e20070ebd.

4. Kelley Dixon and Vince Gilligan, "Episode 506," *Breaking Bad Insider Podcast*, AMC, August 21, 2012, https://www.stitcher.com/podcast/breaking-bad-insider-podcast/e/38708294.

Chapter X

1. Discovery Communications, Inc., "Breaking Bad Special: Adam's Science Experiment," *MythBusters*, accessed November 9, 2018, https://www.discovery.com/tv-shows/mythbusters/videos/breaking-bad-special-adams-science-experiment/.

2. Anne Marie Helmenstine, "What Is the World's Strongest Superacid?," ThoughtCo., last updated October 6, 2018, https://www.thoughtco.com/the-worlds-strongest-superacid-603639.

3. Ibid.

4. "MythBusters Episode 206: Breaking Bad Special," MythBusters Results, accessed November 13, 2018, https://mythresults.com/breaking-bad-special.

5. Brian Palmer, "Soluble Dilemma: How Long Does It Take to Dissolve a Human Body?," *Slate*, December 10, 2009, http://www.slate.com/articles/news_and_politics/explainer/2009/12/soluble_dilemma.html.

Chapter XI

1. David Spiegel, "Dissociative Fugue," *Merck Manual*, last updated July 2017, https://www.merckmanuals.com/home/mental-health-disorders/dissociative-disorders/dissociative-fugue.

2. Jane E. Brody, "When a Brain Forgets Where Memory Is," *New York Times*, April 17, 2007, http://www.nytimes.com/2007/04/17/health/psychology/17brody.html.

3. Neel Burton, "Dissociative Fugue: The Mystery of Agatha Christie," *Psychology Today*, March 17, 2012, last updated September 6, 2017, https://www.psychologytoday.com/blog/hide-and-seek/201203/dissociative-fugue-the-mystery-agatha-christie.

4. Steve Bressert, "Dissociate Fugue Symptoms," *PsychCentral*, last updated August 24, 2017, https://psychcentral.com/disorders/dissociative-fugue-symptoms/.

5. Brody, "When a Brain Forgets Where Memory Is."

6. Kelley Dixon and Vince Gilligan, "Episode 509," *Breaking Bad Insider Podcast*, AMC, August 12, 2013, https://www.stitcher.com/podcast/breaking-bad-insider-podcast/e/38708305.

7. Kelley Dixon and Vince Gilligan, "Episode 402," *Breaking Bad Insider Podcast*, AMC, July 26, 2011, https://www.stitcher.com/podcast/breaking-bad-insider-podcast/e/38708223.

8. Kevin Joy, "Panic Attack vs. Anxiety Attack: 6 Things to Know," *Michigan Health*, January 11, 2017, http://healthblog.uofmhealth.org/wellness-prevention/panic-attack-vs-anxiety-attack-6-things-to-know.

9. Cathy Frank, "What Is the Difference Between a Panic Attack and an Anxiety Attack?," *ABC News*, April 15, 2008, http://abcnews.go.com/Health/AnxietyOverview/story?id=4659738.

10. "What Is Anxiety?," Anxiety BC, accessed November 9, 2018, https://www.anxietybc.com/sites/default/files/What_is_Anxiety.pdf.

11. "Understand the Facts: Panic Disorder," Anxiety and Depression Association of America, accessed October 28, 2018, https://adaa.org/understanding-anxiety/panic-disorder-agoraphobia.

12. "Post-Traumatic Stress Disorder (PTSD)," Mayo Clinic, July 6, 2018, http://www.mayoclinic.org/diseases-conditions/post-traumatic-stress-disorder/diagnosis-treatment/diagnosis/dxc-20308556.

13. Ibid.

14. Adam C. Adler, "General Anesthesia," Medscape, last updated June 7, 2018, http://emedicine.medscape.com/article/1271543-overview.

15. George Bimmerle, "'Truth' Drugs in Interrogation," CIA, September 22, 1993, https://www.cia.gov/library/center-for-the-study-of-intelligence/kent-csi/vol5no2/html/v05i2a09p_0001.htm.

Chapter XII

1. George Rush, "It's Not an Act—I Really Do Have Cerebral Palsy, Says Young Star of *Breaking Bad ... ," Daily Mail,* July 6, 2013, http://www.dailymail.co.uk/health/article-2357324/Its-act--I-really-cerebral-palsy-says-young-star-Breaking-Bad--unlike-character-R-J-Mitte-cope-crutches-disability-growing-army-female-fans.html.

2. Ivan Blumenthal, "Cerebral Palsy—Medicolegal Aspects," *Journal of the Royal Society of Medicine* 94, no. 12 (December 2001): 624–627, https://www.ncbi.nlm.nih.gov/pmc/articles/PMC1282294/.

3. "What Is Cerebral Palsy?," Cerebral Palsy Alliance Research Foundation, October 28, 2018, https://www.cerebralpalsy.org.au/what-is-cerebral-palsy/.

4. Ibid.

5. Ibid.

6. Ibid.

7. Kelley Dixon and Vince Gilligan, "Episode 11," *Breaking Bad Insider Podcast,* AMC, May 18, 2009, https://www.stitcher.com/podcast/breaking-bad-insider-podcast/e/38708152.

8. "RJ Mitte, New UCP Celebrity Ambassador," United Cerebral Palsy of the North Bay, accessed October 28, 2018, http://ucpnb.org/ucp-national-conference-awards/2011-ucp-national-conference/young-benefactors-and-rj-mitte/.

9. "RJ Mitte," Keppler Speakers, accessed October 28, 2018, https://www.kepplerspeakers.com/speakers/rj-mitte/speech-topics.

10. "Huntington's Disease Information Page," National Institute of Neurological Disorders and Stroke, last updated June 15, 2018, https://www.ninds.nih.gov/Disorders/All-Disorders/Huntingtons-Disease-Information-Page/.

11. "About Huntington's Disease and Related Disorders," Psychiatry and Behavioral Sciences, Huntington's Disease Center, Johns Hopkins Medicine, accessed October 28, 2018, http://www.hopkinsmedicine.org/psychiatry/specialty_areas/huntingtons_disease/patient_family_resources/education_whatis.html.

12. Kelley Dixon and Vince Gilligan, "Episode 410," *Breaking Bad Insider Podcast*, AMC, September 20, 2011, https://www.stitcher.com/podcast/breaking-bad-insider-podcast/e/38708261.

13. Lacie Glover, "How Much Does It Cost to Have a Baby?," *NerdWallet*, February 27, 2017, https://www.nerdwallet.com/blog/health/medical-costs/how-much-does-it-cost-to-have-a-baby/.

14. "Low Amniotic Fluid Levels: Oligohydramnios: Causes, Risks and Treatment," American Pregnancy Association, last updated May 26, 2017, http://americanpregnancy.org/pregnancy-complications/oligohydramnios/.

15. "Smoking During Pregnancy," Centers for Disease Control and Prevention, last updated February 6, 2018, https://www.cdc.gov/tobacco/basic_information/health_effects/pregnancy/index.htm.

Chapter XIII

1. World Health Organization, "Noncommunicable Diseases—Cancer," in *Health in 2015: From MGDs to SDGs* (Geneva, Switzerland: WHO Press, 2015), 142–143, http://www.who.int/gho/publications/mdgs-sdgs/MDGs-SDGs2015_chapter6_snapshot_cancer.pdf.

2. "What Is Small Cell Lung Cancer?," American Cancer Society, last updated May 16, 2016, https://www.cancer.org/cancer/small-cell-lung-cancer/about/what-is-small-cell-lung-cancer.html.

3. "Lung Cancer Stages," Cancer Treatment Centers of America, last updated February 22, 2017, http://www.cancercenter.com/lung-cancer/stages/tab/non-small-cell-lung-cancer-stage-3.

4. "What Is Non-Small Cell Lung Cancer?," American Cancer Society, last updated May 16, 2016, https://www.cancer.org/cancer/non-small-cell-lung-cancer/about/what-is-non-small-cell-lung-cancer.html.

5. "Non-Small Cell Lung Cancer Risk Factors," American Cancer Society, last updated May 16, 2016, https://www.cancer.org/cancer/non-small-cell-lung-cancer/causes-risks-prevention/risk-factors.html.

6. "Lung Cancer Symptoms: What You Should Know," Cancer Treatment Centers of America, last updated October 8, 2018, http://www.cancercenter.com/lung-cancer/symptoms/.

7. Ibid.

8. Ibid.

9. Kelley Dixon and Vince Gilligan, "Episode 511," *Breaking Bad Insider Podcast*, AMC, August 26, 2013, https://www.stitcher.com/podcast/breaking-bad-insider-podcast/e/38708314.

10. Kelley Dixon and Vince Gilligan, "Episode 509," *Breaking Bad Insider Podcast*, AMC, August 12, 2013, https://www.stitcher.com/podcast/breaking-bad-insider-podcast/e/38708305.

11. Kelley Dixon and Vince Gilligan, "Episode 304," *Breaking Bad Insider Podcast*, AMC, April 13, 2010, https://www.stitcher.com/podcast/breaking-bad-insider-podcast/e/38708175.

12. "Who Are Radiologic Technologists?," American Society of Radiologic Technologists, accessed October 28, 2018, https://www.asrt.org/main/careers/careers-in-radiologic-technology/who-are-radiologic-technologists.

13. *The Nobel Prize in Chemistry 1985*, NobelPrize.org, Nobel Media AB 2018, accessed October 29, 2018, http://www.nobelprize.org/nobel_prizes/chemistry/laureates/1985/.

Chapter XIV

1. "Highest-Rated TV Series (Ever)," *Guinness World Records*, July 15, 2012, http://www.guinnessworldrecords.com/world-records/107604-highest-rated-tv-series-ever.

2. Thomas C. Arnold, "Shellfish Toxicity Treatment & Management," Medscape, December 28, 2015, http://emedicine.medscape.com/article/818505-treatment.

3. "Canadian Poisonous Plants Information System: Lily-of-the-Valley (Common Name)," Canadian Biodiversity Information Facility, last updated June 5, 2013, http://www.cbif.gc.ca/eng/species-bank/canadian-poisonous-plants-information-system/all-plants-common-name/lily-of-the-valley/?id=1370403267143.

4. Fredericka Brown and Kenneth R. Diller, "Calculating the Optimum Temperature for Serving Hot Beverages," *Burns* 34, no. 5 (August 2008): 648–654, https://www.burnsjournal.com/article/S0305-4179(07)00255-0/fulltext.

5. Jennifer Lai, "Poisoning for Dummies: How Much Skill Does It Take to Brew Up a Batch of Ricin?" *Slate*, April 18, 2013, https://slate.com/news-and-politics/2013/04/how-to-make-ricin-you-dont-have-to-be-a-skilled-terrorist.html.

6. "Ricin," *Breaking Bad* Wiki, accessed October 28, 2018, http://breakingbad.wikia.com/wiki/Ricin.

7. "Ricin Toxin from *Ricinus communis* (Castor Beans)," Emergency Preparedness and Response, Centers for Disease Control and Prevention, last updated November 18, 2015, https://emergency.cdc.gov/agent/ricin/facts.asp.

8. Ibid.

9. Kelley Dixon and Vince Gilligan, "Episode 410," *Breaking Bad Insider Podcast*, AMC, September 20, 2011, https://www.stitcher.com/podcast/breaking-bad-insider-podcast/e/38708261.

10. Ibid.

Chapter XV

1. Sara B. Taylor, Candace R. Lewis, and M. Foster Olive, "The Neurocircuitry of Illicit Psychostimulant Addiction: Acute and Chronic Effects in Humans," *Substance Abuse and Rehabilitation* 4 (February 2013): 29–43, https://www.ncbi.nlm.nih.gov/pmc/articles/PMC3931688/.

2. "Methamphetamine," last updated August 10, 2017, https://www.drugs.com/methamphetamine.html.

3. Steven J. Shoptaw, Uyen Kao, and Walter Ling, "Treatment for Amphetamine Psychosis," Cochrane Systematic Review, January 21, 2009, https://www.cochranelibrary.com/cdsr/doi/10.1002/14651858.CD003026.pub3/abstract.

4. "Methamphetamine," last updated August 10, 2017, https://www.drugs.com/methamphetamine.html.

5. Christ Roberts, "Video: Meet the 'San Francisco Meth Zombies,'" NBC, September 5, 2013, https://www.nbcbayarea.com/news/local/Meet-The-San-Francisco-Meth-Zombies-222592881.html.

6. "Crystal Meth: Some Hard Facts about a Hard Drug," Klean Treatment Centers, October 28, 2018, https://kleantreatmentcenters.com/addiction-info/crystal-meth/.

7. Ingrid A. Binswanger, Carolyn Nowels, Karen F. Corsi, Jason Glanz, Jeremy Long, Robert E. Booth, and John F. Steiner, "Return to Drug Use and Overdose after Release from Prison: A Qualitative Study of Risk and Protective Factors," *Addiction Science & Clinical Practice* 7, no. 1 (March 2012): 3, https://www.ncbi.nlm.nih.gov/pmc/articles/PMC3414824/.

8. Kelley Dixon and Vince Gilligan, "Episode 12," *Breaking Bad Insider Podcast*, AMC, May 25, 2009, https://www.stitcher.com/podcast/breaking-bad-insider-podcast/e/38708153.

Chapter XVI

1. *Compilation of EPA Mixing Zone Documents*, United States Environmental Protection Agency, July 2006, https://nepis.epa.gov/Exe/ZyNET.exe/P1004SMI.TXT?ZyActionD=ZyDocument&Client=EPA&Index=2006+Thru+2010&Docs=&Query=&Time=&EndTime=&SearchMethod=1&TocRestrict=n&Toc=&TocEntry=&QField=&QFieldYear=&QFieldMonth=&QFieldDay=&IntQFieldOp=0&ExtQFieldOp=0&XmlQuery=&File=D%3A%5Czyfiles%5CIndex%20Data%5C06thru10%5CTxt%5C00000009%5CP1004SMI.txt&User=ANONYMOUS&Password=anonymous&SortMethod=h%7C-&MaximumDocuments=1&FuzzyDegree=0&ImageQuality=r75g8/r75g8/x150y150g16/i425&Display=hpfr&DefSeekPage=x&SearchBack=ZyActionL&Back=ZyActionS&BackDesc=Results%20page&MaximumPages=1&ZyEntry=1&SeekPage=x&ZyPURL.

2. Kelley Dixon and Vince Gilligan, "Episode 505," *Breaking Bad Insider Podcast*, AMC, August 14, 2012, https://www.stitcher.com/podcast/breaking-bad-insider-podcast/e/38708291.

Chapter XVII

1. Tim Åström and Nargiza Sadyrova, *Data Analysis: Evaluation of an Analytical Procedure*, Stockholm University, September 17, 2017, https://www.researchgate.net/profile/Tim_Astroem2/publication/320189602_Data_analysis_-_Evaluation_of_an_analytic_procedure/links/59d3c0214585150177f96a7f/Data-analysis-Evaluation-of-an-analytic-procedure.

2. "Research Gas Chromatograph Series 580," Gow-Mac, accessed October 28, 2018, http://www.gow-mac.com/products/research-gas-chromatograph.

3. Michael DeGeorge, Jr., and John Weber, "Methamphetamine Urine Toxicology: An In-Depth Review," *Practical Pain Management* 12, no. 10, last updated November 30, 2012, https://www.practicalpainmanagement.com/treatments/pharmacological/non-opioids/methamphetamine-urine-toxicology-depth-review/.

4. "The Science of *Breaking Bad*: Box Cutter," *Weak Interactions*, July 20, 2011, https://weakinteractions.wordpress.com/2011/07/20/the-science-of-breaking-bad-box-cutter//.

5. C. Graham Brittain, *Using Melting Point to Determine Purity of Crystalline Solids*, University of Rhode Island Department of Chemistry, May 18, 2009, https://www.chm.uri.edu/mmcgregor/chm228/use_of_melting_point_apparatus.pdf.

6. Ian Moore, Takeo Sakuma, Michael J. Herrera, and David Kuntz, *LC-MS/MS Chiral Separation of 'd' and 'l' Enantiomers of Amphetamine and Methamphetamine:*

Enantiomeric Separation, AB SCIEX, 2013, https://sciex.com/Documents/Applications/RUO-MKT-02-0403_Amphetamine_chiral_final.pdf.

7. Andrea Sella, "Classic Kit: Kjeldahl Flask," *Chemistry World*, April 28, 2008, https://www.chemistryworld.com/opinion/classic-kit-kjeldahl-flask/3004923.article.

8. Andrea Sella, "Classic Kit: Allihn Condenser," *Chemistry World*, April 28, 2010, https://www.chemistryworld.com/opinion/classic-kit-allihn-condenser/3004899.article.

Chapter XVIII

1. "Timeline" of history and spread of meth, *PBS Frontline*, accessed October 28, 2018, http://www.pbs.org/wgbh/pages/frontline/meth/etc/cron.html.

2. Kelley Dixon and Vince Gilligan, "Episode 7," *Breaking Bad Insider Podcast*, AMC, April 20, 2009, https://www.stitcher.com/podcast/breaking-bad-insider-podcast/e/38708135.

3. Kelley Dixon and Vince Gilligan, "Episode 6," *Breaking Bad Insider Podcast*, AMC, April 15, 2009, https://www.stitcher.com/podcast/breaking-bad-insider-podcast/e/38708128.

4. "Breaking Bad Candy (100g Pack)," The Candy Lady of Old Town, Albuquerque, NM, August 21, 2016, https://www.thecandylady.com/product/breaking-bad-candy-100g-pack/.

5. Dixon and Gilligan, "Episode 7."

6. Jason Wallach, "A Comprehensive Guide to Cooking Meth on 'Breaking Bad,'" *Vice*, August 11, 2013, https://www.vice.com/en_us/article/exmg5n/a-comprehensive-guide-to-cooking-meth-on-breaking-bad.

7. Benjamin Breen, "Meiji Meth: The Deep History of Illicit Drugs," *The Appendix*, August 23, 2013, http://theappendix.net/posts/2013/08/how-drugs-get-discovered.

8. U.S. Department of Justice, *Information Brief: Iodine in Methamphetamine Production*, National Drug Intelligence Center, July 2002, https://www.justice.gov/archive/ndic/pubs1/1467/1467p.pdf.

9. Kelley Dixon and Vince Gilligan, "Episode 9," *Breaking Bad Insider Podcast*, AMC, May4, 2009, https://www.stitcher.com/podcast/breaking-bad-insider-podcast/e/38708144.

10. John W. Mitchell, Kathryn S. Hayes, and Eugene G. Lutz, "Kinetic Study of Methylamine Reforming over a Silica-Alumina Catalyst," *Industrial & Engineering Chemistry Research* 33, no. 2 (February 1994): 181–184, http://pubs.acs.org/doi/abs/10.1021/ie00026a001.

11. "Sketch of Charles Adolphe Wurtz," *Popular Science Monthly* 22 (November 1882), last updated October 2, 2018, https://en.wikisource.org/wiki/Popular_Science_Monthly/Volume_22/November_1882/Sketch_of_Charles_Adolphe_Wurtz; Alan J. Rocke, *The Quiet Revolution: Hermann Kolbe and the Science of Organic Chemistry* (Berkeley: University of California Press, 1993), http://publishing.cdlib.org/ucpressebooks/view?docId=ft5g500723&chunk.id=d0e2385&toc.id=d0e2146&brand=ucpress.

12. Kelley Dixon and Vince Gilligan, "Episode 305," *Breaking Bad Insider Podcast*, AMC, April 20, 2010, https://www.stitcher.com/podcast/breaking-bad-insider-podcast/e/38708181.

13. Ibid.

14. Kelley Dixon and Vince Gilligan, "Episode 405," *Breaking Bad Insider Podcast*, AMC, August 16, 2011, https://www.stitcher.com/podcast/breaking-bad-insider-podcast/e/38708240.

15. Kelley Dixon and Vince Gilligan, "Episode 501," *Breaking Bad Insider Podcast*, AMC, July 17, 2012, https://www.stitcher.com/podcast/breaking-bad-insider-podcast/e/38708278.

16. A. Allen and R. Ely, *Review: Synthetic Methods for Amphetamine*, Array BioPharma Inc., accessed October 29, 2018, http://www.nwafs.org/newsletters/SyntheticAmphetamine.pdf.

17. Kelley Dixon and Vince Gilligan, "Episode 503," *Breaking Bad Insider Podcast*, AMC, July 31, 2012, https://www.stitcher.com/podcast/breaking-bad-insider-podcast/e/38708283.

18. Lucy Harvey, "How Crystal Meth Made It into the Smithsonian (Along with Walter White's Porkpie Hat)," *Smithsonian*, November 12, 2015, http://www.smithsonianmag.com/smithsonian-institution/how-crystal-meth-smithsonian-walter-whites-porkpie-hat-180957258/.

19. Dixon and Gilligan, "Episode 503."

20. Ibid.

21. Ibid.

22. Wallach, "A Comprehensive Guide to Cooking Meth on 'Breaking Bad.'"

23. Kelley Dixon and Vince Gilligan, "Episode 510," *Breaking Bad Insider Podcast*, AMC, August 19, 2013, https://www.stitcher.com/podcast/breaking-bad-insider-podcast/e/38708309.

Chapter XIX

1. Kirsten Acuna, "Watch Aaron Paul and Bryan Cranston Read the 'Breaking Bad' Finale Script and Agree It's Perfect," *Business Insider*, November 25, 2013, http://www.businessinsider.com/aaron-paul-bryan-cranston-read-breaking-bad-finale-2013-11.

2. Kelley Dixon and Vince Gilligan, "Episode 516," *Breaking Bad Insider Podcast*, AMC, September 30, 2013, https://www.stitcher.com/podcast/breaking-bad-insider-podcast/e/20301/38708328.

3. Discovery, "Breaking Bad Finale Breakdown | MythBusters," YouTube, August 25, 2015, https://youtu.be/06t_KP7y8Ao.

Index